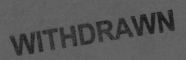

THE
DEPTHS

THE
DEPTHS

The EVOLUTIONARY ORIGINS
of the DEPRESSION EPIDEMIC

JONATHAN ROTTENBERG

BASIC BOOKS

A Member of the Perseus Books Group
New York

Published by Basic Books,
A Member of the Perseus Books Group

Books published by Basic Books are available at special discounts for bulk purchases in the United States by corporations, institutions, and other organizations. For more information, please contact the Special Markets Department at the Perseus Books Group, 2300 Chestnut Street, Suite 200, Philadelphia, PA 19103, or call (800) 810-4145, ext. 5000, or e-mail special.markets@perseus books.com.

Designed by Pauline Brown
Typeset in Stempel Garamond by the Perseus Books Group

Library of Congress Cataloging-in-Publication Data

Rottenberg, Jonathan.
    The depths : the evolutionary origins of the depression epidemic / Jonathan Rottenberg.
        pages   cm
    Includes bibliographical references and index.
    ISBN 978-0-465-02221-2 (hardcover)—
    ISBN 978-0-465-06973-6 (e-book)
    1. Depression, Mental. 2. Depression, Mental—
Treatment. 3. Mood (Psychology) 4. Psychobiology.
5. Evolutionary psychology. I. Title.
    RC537.R6585 2014
    616.85'27—dc23
                                                            2013036462

10 9 8 7 6 5 4 3 2 1

For Laura

Contents

Author's Note

I DON'T BELIEVE IN OBJECTIVITY, THAT A MAN OR WOMAN CAN dispose of his or her biases and have a god's-eye view of a topic. All we can do is our best to be honest and truthful about our motivations.

For my part, I've been on both sides of depression. I've been a depressed subject, wires trailing out of my head, hospital bracelet on my wrist, poked, prodded, questioned about my symptoms. And I've been the scientific objectifier, the one asking the questions, quantifying the behaviors, noting patterns, tabulating the responses into numbers and graphs and ultimately into the currency of journal articles. The experiences of both sides are very different, but each is valid. Each represents *some* truth about this dark, sometimes mysterious topic; each sheds light on our depression epidemic, from different angles. In writing this book, my goal has been to draw both sides together into a complementary synthesis, one that attempts to do justice both to the experiences of patients and to the scientific knowledge we have accumulated in the study of mood and mood disorders.

I am grateful for the people who agreed to be interviewed for this book. I have done my best to relate *their* truth. To protect the identities of interviewees, I have changed names and biographical details throughout.

Jonathan Rottenberg
Tampa, Florida

CHAPTER 1

Why We Need a New Approach to Depression

MORE THAN THIRTY MILLION ADULTS IN THE UNITED STATES SUFFER from depression.[1] Walk down any suburban street in America and start knocking on doors; you'll only need to go five or six houses before finding a resident who bears depression's burden. This is not an American story; you could take the same walk in England, Canada, or Italy with the same results.[2] At the University of South Florida, where I teach abnormal psychology to undergraduates, I recently asked my class: Who among you have been personally affected by serious depression, either in yourselves, in your family, or in a close friend? Seven in ten hands went up. It's impossible to deny: the depressed are our neighbors, our teachers, our doctors, our friends. The depressed are always among us.

Depression's effects ripple out far beyond the affected individual. For the foreseeable future, depression looms as a preeminent public health menace. In a chilling prediction, the

FIGURE 1.1. Change in the Rank Order of Disease Burden for Fourteen Leading Causes Worldwide, 2004–2030.

2004				2030
Disease or injury	**Rank**		**Rank**	**Disease or injury**
Lower respiratory infections	1		1	**Unipolar depressive disorders**
Diarrhoeal diseases	2		2	Ischaemic heart disease
Unipolar depressive disorders	3		3	Road traffic accidents
Ischaemic heart disease	4		4	Cerebrovascular disease
HIV/AIDS	5		5	COPD
Cerebrovascular disease	6		6	Lower respiratory infections
Prematurity and low birth weight	7		7	Hearing loss, adult onset
Birth asphyxia and birth trauma	8		8	Refractive errors
Road traffic accidents	9		9	HIV/AIDS
Neonatal infections and other	10		10	Diabetes mellitus
COPD	13		11	Neonatal infections and other
Refractive errors	14		12	Prematurity and low birth weight
Hearing loss, adult onset	15		15	Birth asphyxia and birth trauma
Diabetes mellitus	19		18	Diarrhoeal diseases

Adapted from *The Global Burden of Disease, 2004 Update*, by the World Health Organization, 2008, Geneva: WHO.

World Health Organization projects that by 2030 the amount of worldwide disability and life lost attributable to depression will be greater than for any other condition, including cancer, stroke, heart disease, accidents, and war (see Figure 1.1).[3] Perhaps most tragically, suicide, an all-too-common outcome of severe depression, now surpasses automotive accidents as a cause of death, with the suicide rate among Americans ages thirty-five to sixty-four increasing by nearly 30 percent just in the last ten years.[4]

This deteriorating situation seems incongruous given the resources we have to combat the noonday demon. There is a growing arsenal of psychological and drug treatments for depression. Social awareness about the symptoms of depression is increasing, and more people are recognizing that it is a bona fide health condition, not a personal weakness or character flaw. Scientific research on depression, from neuroscience to cross-cultural studies, has absolutely exploded.

Yet perversely, as more research and treatment resources have been poured into combating depression, its personal and economic toll has actually grown. Depression now affects more than 15 percent of the population overall, according to our best epidemiological studies,[5] and is striking people at younger and younger ages. A large nationwide survey, the National Comorbidity Survey-Replication, which assessed lifetime depression risk in younger, middle-aged, and older age groups, found that eighteen- to twenty-nine-year-olds are already more likely to have experienced depression than those sixty and older, even though they have been alive for less than half as long.[6] Rampant rates of depression in younger people are worrisome, not only because youth should be a time for blossoming and development, but also because such high rates signal a bleak future for this cohort (see Figure 1.2). Once depression starts, it tends to recur throughout life.

Why, despite all the efforts aimed at understanding, treating, and educating the public about this condition, do rates of depression continue to rise? Why have our treatments plateaued in their effectiveness, and why does the stigma associated with this condition remain very much with us?[7]

Why are we losing the fight against depression?

A Broken Model

Matt had been a straight-A student in high school back in New Jersey. He was a jokester, and beloved by his teachers, who said he was very smart. Now in his second semester at the University of Pennsylvania, he planned to study environmental engineering. He wanted to travel around the world to work on projects in developing countries. But Matt couldn't concentrate. And he was dead tired all the time. Maybe Penn had made a mistake in admitting him? As he walked around campus he saw other students looking down at him. Maybe they were right—he was not as smart, not as rich as them; he was just a chump from Jersey. He retreated to his dorm room. Yes, he was tired and lonely, but he would just bull through; he would do what he had to do to get by.

Yet as the months rolled by, the feelings of fatigue grew. His concentration shot, whenever Matt tried to focus on his course work his mind would go blank, or would drift to thinking about his parents, who had split up one year before. Thoughts of bulling through alternated with thoughts of hopelessness and even fleeting thoughts of ending it all. Somehow Matt made it through all his courses, now earning Cs. Although he got through the year without academic catastrophe, Matt knew something, and something serious, was wrong. It started to dawn on him: "Maybe I'm depressed?"[8]

From the perspective of formal diagnosis, there was no doubt about Matt's condition. He had multiple symptoms of clinical depression. For months he had lost interest or pleasure in things he used to enjoy, experienced crushing fatigue, shown an inability to concentrate, experienced dramatic changes in his sleeping habits, and even had periodic thoughts of death and suicide. These symptoms cast a pall over his freshman year and interfered with his ability to engage with his studies or appreciate the novelty of college life.

FIGURE 1.2. Cumulative Lifetime Prevalence of Major Depressive Disorder by Birth Cohort Among Females.

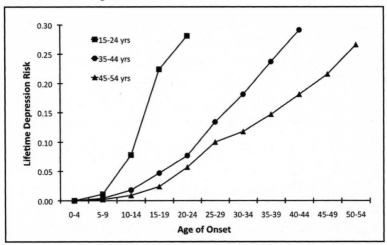

Adapted from data reported in "Sex and Depression in the National Comorbidity Survey. II: Cohort Effects," by R. C. Kessler et al., 1994, *Journal of Affective Disorders, 30*, pp. 15–26.

Matt's symptoms and experiences clearly matched the official category of depression, a major depressive episode, as defined by the American Psychiatric Association's diagnostic manual.

Yet for most sufferers a diagnostic label of depression merely formalizes what they already know, while raising countless other questions. Depression's symptoms are bewildering and disorienting, even after they are properly labeled. Sufferers want to know about the meaning of the symptoms: what they signify, what they represent, and, most of all, why they are happening to them. A diagnosis of depression on its own does not explain the *why*, offers no interpretation of what might be wrong and—as important—what needs to change for all to be set right.

Faced with a case like Matt's, doctors and therapists today invariably assert that the *why* of the symptoms resides
in a deficiency. That deficiency may be in the person's brain
(says the psychiatrist), thoughts (says the cognitive therapist), childhood (psychoanalyst), soul or relationship with
God (priest, pastor, or rabbi),[9] or relationship with a significant other (marital or family therapist).[10] These approaches
appear different on the surface, but all start from the premise
that depression and its symptoms are proof that something
fundamental is wrong.

Because depression is so unpleasant and so impairing,
it may be difficult to imagine that there might be another
way of thinking about it; something this bad *must* be a disease. Yet the defect model causes problems of its own. Some
sufferers avoid getting help because they are leery of being
branded as defective. Others get help and come to believe
what they are repeatedly told in our system of mental health:
that they are deficient.

Depression sufferers thus face two trials. The first is the
depression itself. Its symptoms—despondency, lethargy,
nightly insomnia, an inability to concentrate—are painful
and difficult to manage. The second is facing how others react to the symptoms, hearing the confusing, varied, sometimes hurtful ideas that friends, family, and mental health
professionals posit about what is "wrong with them." Fearing the reactions of others, many conceal their problems and
avoid treatment. The stigma and the impulse to shrink from
depression and depressed people are very much alive. As a
psychiatrist at an inpatient facility put it, "I work in a hospital with 600 beds that has no gift shop; and it has no gift shop

because there isn't the human traffic of people coming to visit people when they're feeling at their worst."[11]

People still feel inclined to whisper when they talk about depression. Depression has no "Race for the Cure"; this condition rarely spawns dance marathons, car washes, or golf tournaments. Consequently, the lacerating pain of depression remains uncomfortably private. One sufferer remarked on her predicament, "[I]t's [depression is] more malignant than cancer. . . . Well I have cancer. I have ovarian cancer and I have severe depression. I'm in five year remission now. When I had cancer, and when I was fighting that, I had flowers, I had people at my door. I had people cooking meals for me. I had people at work, you know rah rah, here we go. When I'm depressed, isolation, people don't call, they don't know what to say, they don't know how to help, they don't know to reach out."[12]

Nearly every depressed person is presented with the idea that his or her underlying problem is a correctible chemical imbalance. We live in a biological age, and this comforting, optimistic notion is popular, embraced by media, patient groups, and mental health professionals. This mindset is supported by the numbers: twenty-seven million Americans take antidepressants.[13] Yet the results often are disappointing. Two-thirds of those treated with antidepressants continue to be burdened with depressive symptoms. Newer antidepressant medications are no more effective than those first developed nearly sixty years ago.

The Star*D treatment trial, one of the largest-ever multi-site treatment studies of the effectiveness of drug therapy for serious clinical depression, found that 72 percent of the 2,876

participants still had significant residual symptoms even after fourteen weeks of antidepressant treatment.[14] These residual symptoms are more than just a nuisance; they include
a nagging low mood, difficulty concentrating, continuing
insomnia, and the feeling that one's self is worthless. These
symptoms are not only debilitating, they are demoralizing.
As Matt put it after two years of taking Lexapro with only
partial improvement, "If the medication can't help me, am I
going to be like this forever?"

Even those patients who initially respond well to a pharmacological treatment are not in the clear. Sadly, their depression will more than likely recur. A major study found
that about half of adolescents who recovered from major depression became depressed again within five years, regardless
of what treatment or therapy they received to get over their
initial depression.[15]

At the current juncture even diehard biological psychiatrists acknowledge that the discovery of a physical cause for
all cases of depression has proven elusive. We have thousands
of biological assays, from brain imaging to blood draws, but
still no biological test for depression. Without a clear target
of what is being treated, the search for a magic pharmacological bullet for depression verges on the quixotic.

A depressed person can also expect to be offered a psychological interpretation of what is wrong with him or her.
For example, cognitive approaches see depression as due to
faulty thinking, a consequence of distorted ideas such as *I'm
a failure, nobody loves me,* or *the future is totally hopeless.*[16]
This approach has spawned an influential treatment called
cognitive-behavioral therapy (CBT), a practice intended to

correct thought. Like the chemical imbalance theory, psychologically based defect models also exaggerate the case. CBT is about as effective as treatment with antidepressants: beneficial for many, but far from a cure.

In fact, it remains murky why CBT works when it does work.[17] Just as aspirin's effectiveness does not prove that headaches are caused by a lack of aspirin, the successes of cognitive therapy don't mean that depression is caused by cognitive defects. Like the search for biological defects, the search for the cognitive defect that *causes* depression hasn't produced clear answers.

Yes, our pharmacologically and psychologically based treatments are better than nothing. But unfortunately these conventional approaches are far from cures for most sufferers. And ironically, just as the public has become more accepting of seeking treatment for depression, it is not aware of the modest effectiveness of available options. Only recently have there been signs that this is starting to change. A new analysis of six major clinical trials in the *Journal of the American Medical Association* found that common antidepressants worked little better than placebos for people with mild to moderate depression. This report received animated discussion on CNN, in the *New York Times*, and in other major media outlets.[18]

So, why *are* we losing the fight against depression?

I have come to believe that the intuitively appealing idea that depression stems from defects has led us directly to our current impasse. If you go to a conference in clinical psychology or psychiatry, I can promise you will experience two things. One, you will hear many fascinating presentations on the cognitive, social, biological, and developmental aspects of depression. Two, you will be unlikely to hear much about the depression epidemic. This seems odd until you realize that none of the major research paradigms equips us to understand why we are beset by a depression epidemic. If depression results from faulty cognitions, why would our cognitions suddenly become so faulty? If it's faulty biology at work, why would our equipment fail us now, and on a mass scale? Our genetic endowment, for example, does not turn on a dime. Even if one looks to the environment, which is always changing, it's not immediately obvious what aspect of it has changed so drastically as to account for such a surge in depression.

In challenging the depression-as-defect view, it is reasonable to wonder about the alternatives. Some commentators and scholars have gone to the other extreme, arguing that depression is beneficial. From improved problem solving to resource conservation, several accounts put the focus on depression's overlooked benefits. So if we reject the disease model, it seems we must adopt the position that depression is good.

Or must we?

One sufferer implicitly rejected this overly simplistic choice, saying about her depression: "It sucks, but there's value in it."[19] In the pages to come, I hope to show that taking this more nuanced position allows us to ask more interesting questions about depression. Depression is potentially good *and* bad, a point of departure that may help us get closer to the mystery of what depression is, why so many suffer from it, and *why* it is such a tough nut to crack.

The Mood Science Approach

At the center of the nut is mood. Depression's defining feature is persistent low mood. The typical depressed person reports moods that are excessively dull, empty, and sad, as well as moods that lack joy, excitement, or cheer. The centrality of mood to depression is reflected in its classification as a mood disorder.[20]

Yet modern approaches to depression—be they biological, cognitive, or social—have focused on just about everything *but* mood. In part, this is because the study of mood had little momentum for most of the twentieth century. Researchers had little interest in the topic; skeptics questioned whether something as evanescent as mood could ever be studied with precision or objectivity.

Just as CAT scans and functional magnetic imaging allowed physicians to see the innermost recesses of the body, so, too, in the last thirty years an increasingly sophisticated set of tools has enabled us to measure mood and emotion. The emerging field known as *affective science* now benefits from an enviable wealth of measurement tools, with standard techniques for measuring the moods that people report; systems for measuring behavior in the lab and in the field; and new ways to monitor the physiology of mood and emotion, from functional brain scans to miniature sensors that monitor the body as people go about their everyday lives.

Amid these exciting developments in the mid-1990s, I arrived at Stanford University as a new graduate student in psychology, full of hope and naiveté. There I saw other scientists beginning to apply the methods and insights from affective science to the study of psychopathology. Ann Kring at Berkeley, one of my idols in the field, was using these techniques to discern how schizophrenia altered feelings and emotional behaviors, observations that cast the disorder in an entirely new light.[21] As I watched Ann give a presentation on her work at Stanford, I thought, "We need to do this for depression!"

I am no longer that bright-eyed student, but, in the years since those California days, it has become increasingly clear to me that affective science holds the key to understanding and treating depression. And as the depression epidemic has accelerated, getting at its root causes has become a matter of some urgency. This book is above all an attempt to elucidate the relationship between mood and depression. Our model is broken. We need to usher in a new diagnostic and therapeutic paradigm, one based in the science of mood.

———

To appreciate what affective science can tell us about mood disorders, we first need to understand what moods are. Why do we have them at all? Here we explore the architecture of the mood system, an ancient system that influences what we feel, think, and do, as well as guiding our bodily responses to the world.

All organisms—from planaria to sidewinders to rock stars—face the great problem of behavior. What, given a limitless menu of possibilities, should a creature do? A billy goat by the farmhouse can eat a tin can, take a nap, chase chickens, or run in circles. How does it decide what to do first? Fortunately the goat, like all the animals on the farm, has a head start on this problem, because it has been equipped with a behavioral guidance system that moves it toward actions that have been successful in the past (which is to say, actions that led ancestor goats to successfully reproduce and spread their genes). In other words, moods are internal signals that motivate behavior and move it in the right direction. To understand the formidable role that moods play in survival, remember Charles Darwin's theory of evolution and his profound idea that evolutionary pressures shaped not only physical features but animals' mental processes and behavioral characteristics as well.

As a first step, the mood system needs to know what kind of situation it is in. Different situations have different implications for fitness (i.e., survival and reproduction). For our goat, the situation includes the external world of the barnyard: Is it dark or high noon? Hot or cold? Is food nearby and plentiful, or far away and scarce? Might there be predators about? The situation also includes the goat's internal world: Is it bleeding, sick, or in pain? Hungry or satiated? All of these elements affect mood. The mood system, then, is the great integrator. It takes in information about the external and internal worlds and summarizes what is favorable or unfavorable in terms of accomplishing key goals related to survival and reproduction.[22]

These computations are automatic. The goat is unaware it is doing evolution's bidding when it eats a carrot. Eating a carrot feels good for a reason: animals feel pleasure when they pursue actions that lead to survival and reproduction.[23] Moods sculpt behaviors in ways that enhance fitness and do so without the animal's express permission or knowledge.

Yet moods are more than a summary readout of the status quo—they set the stage for specific emotional behaviors. Most of us have experienced a situation in which an irritable mood made it easier for a minor slight to trigger an outburst of rage, or when an anxious mood made us so jumpy that just a few strange noises in the night provoked full panic and terror.[24] Confirming scientists' intuitions, controlled experiments find that an anxious mood narrows the focus of attention to threats. When anxious subjects are shown happy, neutral, and angry faces on a computer screen, their attention is drawn to the angry faces signaling a potential threat.[25] Conversely, good moods broaden attention and make people inclined to seek out information and novelty.[26] In one study, participants in good moods sought more variety when choosing among packaged foods, such as crackers, soup, and snacks.[27] Moods have the power to influence behavior because they have such wide purchase on the body and mind. They affect what we notice, our levels of alertness and energy,[28] and what goals we choose.

Finally, once a goal is embarked upon, the mood system monitors progress toward its attainment. It will redouble effort when minor obstacles arise. If progress stops entirely because of an insuperable obstacle, the mood system puts the brakes on effort.[29] Experiments have successfully tested the idea that negative mood mobilizes effort when tasks become challenging. When participants are put in a negative mood and subsequently are given a difficult task to perform, they can be expected to show a larger spike in blood pressure, a key index of bodily mobilization. Yet if the task is made significantly more difficult, to the point that success is no longer possible, participants no longer demonstrate the sharp spike, a sign that the mood system de-escalates effort for impossible (or seemingly impossible) tasks.[30]

The switch makes sense. Given that nearly all key resources are finite (be they time, energy, or money), expending them on unreachable goals can be ruinous. This is particularly evident in goals related to physical survival, such as food seeking. When a bear catches no salmon after hours of working a favorite bend in the river, the mood system decides that it's time to pull back and move on. The same principle also applies to longer-term commitments. Take the goal of bearing a child, a deeply held commitment for many women. We could expect that for a woman who has such a goal and has not yet fulfilled it, menopause would be accompanied by a period of low mood that would eventually diminish after she gives up on this now-unreachable goal and adjusts to reality. We would also expect that a woman who continued to want to bear a child despite its impossibility would experience a further escalation of low mood. Research supports these predictions exactly.[31]

Mood flexibly tunes behavior to situational requirements, which is what makes it so effective as an adaptation. When a situation is favorable, high moods lead to more efficient pursuit of rewards. Reward-seeking behavior is invigorated (eat grass while the sun shines). In an unfavorable situation, low moods focus attention on threats and obstacles, and behavior is pulled back (hunker down until the blizzard ends).[32] Mood reflects the availability of key resources in the environment, both external (food, allies, potential mates) and internal (fatigue, hormone levels, adequacy of hydration), and ensures that an animal does not waste precious time and energy on fruitless or even dangerous efforts (doing a mating dance when predators are lurking).

More Than Words

One of the amazing things about the mood system is how much of it operates outside of conscious awareness. Moods, like most adaptations, developed in species that had neither language nor culture.[33] Yet words are the first things that come to mind when most people think about moods. We are "mad," we are "sad," we are "glad." So infatuated are we with language that both laypeople and scientists find it tempting to equate the language we use to describe mood with mood itself.

This is a big mistake. We need to shed this language-centric view of mood, even if it threatens our pride to accept that we share a fundamental element of our mental toolkit with rabbits and roadrunners. Holding to a myth of human uniqueness puts us in an untenable position. For one thing, it would mean that we deny mood to those humans who have not yet acquired mood language (babies) or have lost mood language (Alzheimer's patients). Toddlers, goats, and chimps all lack the words to describe the internal signals that track their efforts to find a mate, food, or a new ally; their moods can shape behavior without being named.[34] Language is not required for moods. All that is needed is some capability for wakeful alertness and conscious perception, including the perception of pain and pleasure, which is certainly present in all mammals.[35]

Further, relying solely on language provides a misleading picture of what moods are really all about. Although a sad mood involves states we might label as "down" or "depressed," moods encompass the full body and mind, from drooped posture and downcast glances to changes in immune and hormonal systems and darkened perception and memory (we notice every slight, every fault, and are flooded with memories of past failure).[36] It is telling that severely depressed humans find verbal labels like "sad" or "down" pitifully inadequate to describe their inner sensations and experiences.[37] What we say about our feelings is only one window on mood. Because mood leaves more than one kind of fingerprint,[38] we need to be open to a variety of evidence—in the mind, in the brain, and in behavior—to appreciate moods in action.

Mesmerized by our linguistic abilities, it is understandable that humans feel compelled to tell ourselves stories about our moods. Moods, especially intense moods, by their nature grab attention and call for explanation. Next time you are in a brooding, seething stew of an irritable mood, see if you can resist the urge to explain *why*.

Yet despite this impulse, the stories we tell ourselves about our moods are fraught with error. We hypothesize that we feel down because we have gotten behind at work; the true reason for the feeling may be that we are getting over a cold and our bodies are depleted of strength. At other times, try as we might, we cannot generate any story for our mood (*I don't know why, I just feel low*). We are forlorn *and* baffled. At a loss, we might turn to a therapist to help us revise our story.[39]

Of course making sense of our feelings is not always a hopeless task. If a driver cuts us off in traffic, we know full well why we are suddenly balling our fists and yelling. This burst of anger is the hallmark of an emotion, defined as a short-term reaction to a specific event. So, too, with other emotions; if we have a sudden burst of fear, or of embarrassment, we usually have a story at the ready: the big hairy spider, the glass of red wine that has spilled on our lap.

Moods are different. Moods take longer to come on and to go away. They are an overall summary of the various cues around us. And usually they are harder to sort out. Because humans operate in complex environments that contain a confusing buzz of ever-changing objects, getting a fix on our moods is more challenging than it seems.[40] Our heavy reliance on symbolic representation also makes the precipitants of low mood more idiosyncratic in our species than in others. We become sad because Bambi's mother dies, because there are starving people a continent away, because of a factory closing, because of a World Series defeat in extra innings. Though there is a core theme of loss that cuts across species, humans' capacity for language enables a larger number of objects to enter, and alter, the mood system.

Despite our deep yearning to explicate moods, the average person cannot see many of the most important influences on mood. As the great integrator, the mood system is acted on by many potential objects, and many of the forces that act on mood are hidden from conscious awareness (such as stress hormones or the state of our immune system). Left to our own devices, the stories we tell ourselves about our moods often end up being just that. Stories.

That's where mood science comes in.

Fortunately, a systematic, research-based mood science approach has begun to replace folk wisdom (or folk ignorance) about mood with hard data. Although our ability to predict mood in a specific person is not yet as accurate as tomorrow's weather prediction, a growing body of work is starting to reveal the many factors that influence mood, from inborn temperaments to transient events to daily routines. One of the main strengths of mood science—particularly useful for the purposes of this book—is that the same factors can be used to explain both typical mood variation and extreme moods like severe depression. The mood science approach thus has unique potential for explaining why we are in the middle of a depression epidemic.

———

WE MUST UNDERSTAND the ultimate sources of depression if we are ever to get it under control. To do so, we need to step back and replace the defunct defect model with a completely different approach. The mood science approach will be both historical and integrative: historical because we cannot understand why depressed mood is so prevalent until we understand why we have the capacity for low mood in the first place, and integrative because a host of different forces (many hidden) simultaneously act on people to impel them into the kinds of low moods that breed serious depression. Further, we will also integrate how people *respond* to periods of low mood, including responses that (even with the best of intentions) often have the paradoxical effect of making depression worse.

Stepping back means that *The Depths* has an immodestly large scope, spanning the ultimate origins of the capacity for depression to the forces that impel people in and draw people out of depressive episodes. Although it might be comforting to blame someone or something, no single villain or cause can explain the entire depression epidemic. Nor is there a single factor that, if changed, would reverse the epidemic.

Instead of proposing yet another single-bullet theory of depression, the chapters ahead detail a remarkable confluence of unfortunate circumstances. Some began many millions of years ago and are built into the architecture of our mood system, whereas others, like human language, are of more recent advent, and still others reflect cultural and social factors operating in the last twenty or thirty years. By examining these circumstances, we can begin to understand how together they have created the perfect storm of mood. Only then will we get to the bottom of the depths of depression—and in so doing, discover new ways to climb back out.

CHAPTER 2

Where the
Depths Begin

OUR BODIES ARE A COLLECTION OF ADAPTATIONS, EVOLUTIONARY
legacies that have helped us survive and reproduce in the face
of uncertainty and risk. That does not mean that adaptations
are perfect; far from it. Evolutionary thinkers have long cau-
tioned against thinking of adaptations as inevitable steps up a
ladder of progress, conferring ever greater benefits. Flawed
designs, if they promote survival and reproduction, are more
than good enough.

Hence we should expect that even the most wondrous
adaptations come with costs. The evolution of bigger brains
in humans not only enabled higher cognitive ability but
also increased the risks of childbirth. The advent of bipedal
walking freed up our hands for improved hunting and crafts-
manship, but at the same time upright posture placed new
pressures on the spinal column, rendering our species prone
to back injuries and pain. The same cost-benefit calculus

FIGURE 2.1. Warm Blood Is Usually a Benefit, Despite the High Metabolic Demands That It Imposes on Mammalian Species.

Photo credit: Kev Chapman

holds as we look across the animal kingdom. Most mammals evolved to be endothermic, or warm blooded, because this trait allowed them to forage and hunt in cold weather (see Figure 2.1), unlike their reptilian competitors. Although the benefits are obvious, keeping blood warm exacts a big cost: mammals must eat more food than most reptiles or risk malnutrition or starvation.[1]

The cost-benefit calculus applies to psychological adaptations as well. The layperson might assume that high moods are always good and low moods are always bad. Not so. Both present pluses and minuses. We are born with the capacity for both high and low moods because each has, on average, presented more fitness benefits than costs. Just as being warm blooded can be a liability, high moods are increasingly understood as having a "dark side," sometimes enabling rash, impulsive, and even destructive behavior.[2] Likewise the capacity for low mood is accompanied by a bundle of benefits and costs. Seen this way, depression follows our adaptation for low mood like a shadow—it's an inevitable outcome of a natural process, neither wholly good nor entirely bad.

Rather than diving into why depression exists, we should begin with a simpler investigation. What evolutionary advantages does low mood confer? Why does it persist, despite what might look like an awful risk of plunging an organism into depression?[3]

Benefits of Low Mood

Ever since Charles Darwin saw signs of dejection in orangutans and chimps,[4] the behavioral sciences have launched a raft of theories about the adaptive value of low mood. One theory starts from the premise that because confrontations are a common and dangerous consequence of competition, low mood helps de-escalate conflicts. By helping the loser to yield, low mood allows him or her to live to fight another day. Another theory highlights the value of low mood as a "stop mechanism," a means of discouraging effort in situations in which persisting in a goal is likely to be wasteful or dangerous. Still another theory proposes that low mood states help sensitize people to "social risk" and help them reconnect when they are on the verge of being excluded from a group. And yet another theory suggests that low mood is adaptive because it enables people to make better analyses of their environments, which could be especially useful when they are facing difficult problems.[5]

At first blush, the existence of multiple theories seems problematic. How can we decide which is right? Upon closer inspection, however, it becomes apparent that these theories are trains that run on parallel tracks. Each theory helps explain part of why low mood would be conserved over evolutionary time. Although none is sufficient by itself, when the theories are arrayed together, we can begin to appreciate why low mood endures: it is a state that is potentially useful in many different situations.

Of course it is very possible that some theories are more right than others. Moreover, theories on their own prove little. One of the main challenges in building a convincing case for particular functions of low mood is to show that the putative benefits are more than a theoretical proposition. Fortunately hard data from well-controlled experiments support some of the functions of low mood that have been proposed.[6]

One idea that has been repeatedly tested is that low mood can make people better at analyzing their environments. Classic experiments by psychologists Lyn Abramson and Lauren Alloy focused specifically on the accuracy of people's perceptions of their control of events, using test situations that systematically varied in how much control the subject truly had. In different conditions, subjects' responses (pressing or not pressing a button) controlled an environmental outcome (turning on a green light) to varying degrees. Interestingly, subjects who were dysphoric (in a negative mood and exhibiting other symptoms of depression) were superior at this task to subjects who were nondysphoric (in a normal mood). Subjects who were in a normal mood were more likely to overestimate or underestimate how much control they had over the light coming on.[7]

Dubbed *depressive realism*, Alloy and Abramson's work has inspired other, often quite sophisticated, experimental

FIGURE 2.2. Negative Mood Enhances the Quality and Concreteness of Persuasive Arguments.

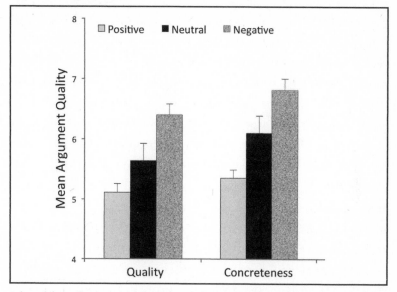

Adapted from data reported in "When Sad Is Better Than Happy: Negative Affect Can Improve the Quality and Effectiveness of Persuasive Messages and Social Influence Strategies," by J. P. Forgas, 2007, *Journal of Experimental Social Psychology*, *43*, pp. 513–528.

demonstrations of ways that low mood can lead to better, clearer thinking.[8] In 2007 studies by Australian psychologist Joseph Forgas found that a brief mood induction changed how well people were able to argue. Compared to subjects in a positive mood, subjects who were put in a negative mood (by watching a ten-minute film about death from cancer) produced more effective persuasive messages on a standardized topic such as raising student fees or aboriginal land rights. Follow-up analyses found that the key reason the sadder people were more persuasive was that their arguments were richer in concrete detail (see Figure 2.2).[9] In other experiments, Forgas and his colleagues have demonstrated diverse benefits of a sad mood. It can improve memory performance, reduce errors in judgment, make people slightly better at detecting deception in others, and foster more effective interpersonal strategies, such as increasing the politeness of requests. What seems to tie together these disparate effects is that a sad mood, at least of the garden variety, makes people more deliberate, skeptical, and careful in how they process information from their environment.[10]

It is not surprising that the provocative hypothesis of depressive realism has also been subject to attack, and systematic efforts to pin down exactly when it is likely to be observed continue.[11] Yet that sad mood *ever* enhances cognitive function should make one stop to ponder what exactly we mean by "normal" mood. If people who are in a sad mood sometimes assess the world quite accurately, people in a "normal," healthy mood may be less in touch with reality. At least some data suggest that people in a normal mood can be prone to positive illusions, overconfidence, and blindness to faults.[12]

Arguing about the functions of mood can be challenging. Some hypothesized functions of mood play out over time and are nearly impossible to test decisively with a laboratory experiment. Take the hypotheses that (1) low mood helps people disengage from unattainable goals and (2) we end up better off as a result of letting go. Testing this hypothesized chain of events requires data about the real-world goals that people want to attain and the ability to measure people's adjustment and well-being over the longer term. A nonexperimental study of adolescent girls in Canada did just this, collecting four waves of longitudinal data on the relationship between goals and depression over nineteen months. Consistent with the first hypothesis, those adolescents who had depressive symptoms reported a tendency to become more disengaged from goals over time. The stereotypical image of a disengaged adolescent sulking in her room with an iPod may not look like the process of rebuilding psychological health. Results were in fact consistent with the idea that letting go was a positive development: those adolescents who became more disengaged from goals ended up being better off, reporting lower levels of depression in the later assessments.[13]

As data accumulate to support the benefits of low mood, we shouldn't be surprised that it is good for more than one thing. Multiple utilities are the hallmark of an adaptation. We see this elsewhere in the body. Take, for example, eyelids. Closing our eyes protects them from damage from foreign bodies or overly bright light. Blinking every few seconds moves tears over the cornea, keeping it moist. Keeping the eyelids closed during sleep protects the eye and prevents dryness. Eyelids enhance fitness because they are good for many things.

The idea that low mood could have more than one function squares with the obvious fact that it is triggered reliably by very different situations. A partial list of triggers includes separation from the group, removal to an unfamiliar environment, the inability to escape from a stressful situation, death of a significant other,[14] scarce food resources, prolonged bodily pain, and social defeat.[15]

In humans the value of low mood is put to the fullest test when people face serious situations in which immediate problems need to be carefully assessed. We might think of the groom who is left at the altar, the loyal employee who is suddenly fired from his job, or the death of a child. If we had to find a unifying function for low mood across these diverse situations, it would be that of an emotional cocoon, a space to pause and analyze what has gone wrong. In this mode, we will stop what we are doing, assess the situation, draw in others, and, if necessary, change course.

Fantasizing about a world without low mood is a vain exercise. Low moods have existed in some form across human cultures for many thousands of years.[16] One way to appreciate why these states have enduring value is to ponder what would happen if we had no capacity for them. Just as animals with no capacity for anxiety were gobbled up by predators long ago, without the capacity for sadness, we and other animals would probably commit rash acts and repeat costly mistakes. Physical pain teaches a child to avoid hot burners; psychic pain teaches us to navigate life's rocky shoals with due caution.[17]

Writer Lee Stringer, reflecting on his serious depression, put this idea in far more poetic terms: "Perhaps what we call depression isn't really a disorder at all but, like physical pain, an alarm of sorts, alerting us that something is undoubtedly wrong; that perhaps it is time to stop, take a time-out, take as long as it takes, and attend to the unaddressed business of filling our souls."[18] Stringer's experience reminds us that the unpleasant or even unattractive aspects of low mood are not necessarily at odds with its utility. People in a low mood blame and criticize themselves, repeatedly turn over in their heads situations that went wrong, and are pessimistic about the future. These characteristics, although uncomfortable, are also potentially useful. A keen awareness of what has already gone wrong and what can do so again can help a person avoid similar stressors in the future. In Randolph Nesse's elegant phrase, these features of low mood "can prevent calamity even while they perpetuate misery."[19]

Costs of Low Mood

Low mood's potential benefits help explain why it has endured. But we should be skeptical of any theory that claims a trait is always useful or adaptive. Periods of low mood potentially create vulnerabilities. Among the most salient are behavioral vulnerabilities. Doing nothing can be risky; in times past, prolonged immobility could increase the risk of being eaten by a predator. Or a window of opportunity may close.

There are also potential cognitive vulnerabilities. Severely depressed people are capable of breathtakingly distorted thinking that appears to be the polar opposite of depressive realism. It's not obvious what benefits anyone could receive from psychotic thoughts such as, "I am the devil," "I am guilty of all the world's sins," or "I believe all of my organs are rotting from within."[20]

This distorted thinking can lead to odd, seemingly self-destructive behavior. Dr. Frenk Peeters recalled a severely depressed woman who was referred to his psychiatric group for evaluation. After hearing their professional opinion that she urgently needed help, she acknowledged that she needed care but insisted that she couldn't start treatment because she did not have the money to pay for it. Her statement was curious because it was untrue: her financial situation was good. Yet she continued to refuse treatment because of the delusional belief that she was poor.[21]

Those who suffer from severe depression often complain that they are having trouble thinking. "I feel as if my brain were a lump of protoplasm," begins one vivid description, "with tiny circuits embedded in it, and some of the wires keep shorting out. There are tiny little electrical fires up there, leaving crispy sections of neurons smoking and ruined."[22] There is a term in neuropsychology for this domain, *executive functioning*. Though it may conjure up a vision of a tiny, well-dressed man residing in one's head, this term actually refers to a suite of essential cognitive abilities involving mental control. This includes the ability to keep material alive in working memory (i.e., the names of people you just met at a meeting) as well as to attend to more than one thing simultaneously (giving a presentation and monitoring the expressions of people in the audience for comprehension). Consistent with clinical reports and patients' own impressions, studies have found that serious depression can weaken several aspects of executive functioning.[23] And it is this weakening—marked by an impaired ability to focus and concentrate effectively on a job or schoolwork—that often drives even the most reluctant sufferer into treatment.

We do not yet have a detailed understanding of when and where low mood becomes costly. Few scientists have tried to reconcile the evidence that low mood has both benefits and costs.[24] Most of the debate about depression has been polarized into mutually exclusive depression-is-good versus depression-is-bad camps.[25] The time has come to bring these camps together to a more nuanced position.

Shallow and Deep Depression

Low mood comes in different shades of gray. This fundamental fact applies to all aspects of depression, including the discussion of its costs and benefits. Low mood can last from minutes to years and can be barely noticeable or punishingly severe. For purposes of this discussion, I distinguish between milder periods of low mood, which I call *shallow depression*, and crippling periods of low mood that are both long and strong, which I refer to as *deep depression*. I reserve the latter term for a mood disturbance that exceeds our current diagnostic threshold for a major depressive episode; that is, a mood disturbance that is accompanied by five or more symptoms *and* that lasts for at least two weeks.

One way to try to reconcile low mood's costs and benefits is to focus on severity: shallow depression is adaptive, whereas deep depression is a maladaptive disease. Indeed, critics who reject the idea that low mood has evolutionary utility naturally focus on severe cases of depression: the patient who is flat on his or her back, laid low, and unable to work or go to school.[26] Surely cases like this must represent some disease or defect?

One reason to be skeptical is that sometimes even people with deep depression can outperform healthy people on a cognitive task. For example, in a controlled, sequential decision-making task designed to simulate a real-world hiring decision (choosing a secretary from among a series of applicants), deeply depressed inpatients tended to choose better candidates than both healthy participants and those who were recovering from a depressive episode.[27] Although results like this are rare, they suggest that a reconciliation based on severity is problematic.

There are other reasons that reconciling the adaptive value of low mood based on the severity of the mood is likely to be unworkable. First, it is difficult to isolate what's different about the subgroup of people who have the "depression disease." We return to a glaring problem with defect models: no one has identified the basis of the disease, the underlying defect in the mind or brain that causes deep depression. For example, after a series of false leads, the field of genetics has backed away from approaches that hold single genes responsible for many or most depressions.[28] A similar story could be told for neuroimaging, endocrinology, or cognitive approaches to depression: despite promising suspects, no definitive causes have been identified.

Even if we drop the search for direct causes and hone in on research that focuses on risk factors, the problem of making sharp distinctions between deep and shallow depression remains. *Risk factors* are the variables that raise the probability that an event will occur. For example, age is a risk factor for developing dementia. Research on risk factors for deep depression has revealed many trends. We know that people who lack social support, face high levels of environmental stress, have poor sleep habits, or have a fearful temperament all are more likely to experience deep depression. However, these risk factors do not bring us closer to isolating the disease process that is responsible for deep depression, because these exact same factors also put a person at increased risk for shallow depression.[29] The existence of a common set of risk factors for shallow and deep depression suggests that we are studying one thing, mood, which varies along a continuum of strength. Ignoring this would be like a weather forecaster using separate models to predict warm days and very hot days rather than considering general factors that predict temperature.

Importantly, thinking about mood in a unified way fits with what we know about the epidemiology of depression, specifically with how low mood flows through time. Extensive longitudinal studies conducted on thousands of individuals consistently show that shallow, low-grade depression is a precursor of serious, deep depression. That is to say, more often than not, someone who develops deep, disabling depression will start out with shallow depression.[30] Likewise, in the aftermath of deep depression, even with treatment, it is typical for patients to continue to be bothered by periods of shallow depression (a hangover of symptoms).[31] Longitudinal studies of the week-to-week course of depression also show frequent transitions between shallow and deep depression.[32] Over the course of a depression episode (they last, on average, about six months), a person may experience five or six of these transitions.[33] It makes little sense that every one of these transitions represents a move between an adaptive and a diseased state.

Just as there are not separate adaptations for minor jitters versus paralyzing anxiety, or for mild versus excruciating pain, there is no separate evolutionary explanation for deep depression. Once the capacity for shallow depression evolved, it was inevitable that an intense variant, in the form of deep depression, would appear. I consider the epidemic of deep depression by taking a more unified approach to mood. I'll address two sets of influences on mood: those forces that render so many people vulnerable to long periods of shallow depression, and then those forces that worsen shallow depression.

The Changing Cost-Benefit Ratio of Adaptations

As mentioned at the outset of this chapter, for any adaptation, we must accept the bad with the good. The benefits of an adaptation can be surprisingly fragile. They may, for example, play out only if an animal is in its typical environment. In the dense primeval forest, deer that freeze at the first sniff of a wolf were (and are) less likely to be seen by a predator skilled at detecting movement. Deer evolved to freeze at the first sign of danger. However, we need only think of the deer frozen in the headlights to know that even a generally useful behavior is not useful in all environments. The advent of the motorcar increased the costs associated with deer-freezing behavior, especially for those deer that live in wolf-free suburbs.

An example somewhat closer to home is the human tendency to select and eat calorically dense foods when they are available. This tendency has historically conferred more benefits than costs, because the specter of famine has loomed for nearly all of evolutionary time. The costs associated with efficient storage of food energy and a preference for rich foodstuffs only become apparent in modern environments in which food is abundant and the drive-through McDonald's is ubiquitous. Such characteristics contribute to our obesity epidemic and the rise of obesity-related conditions such as diabetes.[34]

Similar "mismatch" scenarios have also been identified for psychiatric symptoms. It has been proposed that anxiety abounds because our evolved mechanisms for generating anxiety are out of sync with modern triggers. When we want to put the final touches on tomorrow's big sales presentation, a vigorous "fight or flight" response—so good for detecting stalking lions on the savanna—scrambles our thoughts and leaves us too keyed up to sleep.[35] As the triggers for anxiety change, reactions that saved us in the past may drag us down in the present day.

The coming chapters discuss the ways that low mood is akin to other psychological capacities such as anxiety and pain, which are at once important defenses against threats and damage, as well as lurking vulnerabilities in the form of disabling anxiety and pain conditions. If we grant that low mood is an adaptation that is always costly (to some degree),[36] we can ask whether periods of low mood may have become more costly in our contemporary environment.[37]

Although this book is not a work of history, it is worth considering how recent history may have set up inauspicious conditions for mood. In the chapters to come I show that current environmental conditions may exploit vulnerabilities of the mood system. These include the possibilities that typical triggers for low mood have changed (and become harder to resolve) and that our attitudes toward sadness have changed (to less effective responses). Or even the possibility that our expectations about happiness have changed dramatically, and as they rise, ironically, are making low moods harder to bear than ever before.

In considering whether the cost-benefit ratio of low mood has shifted, it is easy to forget both the recent advent of *Homo sapiens* and the breakneck speed of historical change since our species came on the scene, compared to the slow pace of natural selection. Consider that *Homo sapiens* has only been around for a few hundred thousand years, a tiny fraction of the three hundred-million-year tenure of mammals (humans are relative latecomers). On an evolutionary timescale, we are last-minute gate-crashers.

Ultimately it is environmental characteristics that generate selection pressure on traits (i.e., if an ice age comes, creatures who possess cold-tolerant traits will be more likely to survive and pass on their genes). Critically, nearly all of previous human existence took place in a radically different environment than the one we now inhabit. A reasonable estimate is that our species lived (and evolved) as hunter-gatherers one thousand times longer than in any other lifestyle. Although detailed reconstruction of the hunter-gatherer world is impossible, our mood system was surely forged in a context in which life was short (one could expect to live to age thirty) and various existential threats were always at the door, be they starvation, death from disease, predation, or war. Evolutionary psychologists John Tooby and Leda Cosmides sum this up well: "The world that seems so familiar to you and me, a world with roads, schools, grocery stores, factories, farms, and nation-states, has lasted for only an eyeblink of time when compared to our entire evolutionary history."[38]

Although our species *is* still evolving (the capacity to digest milk and malarial resistance are recent), there is no way for evolution to keep pace with the furious and radical changes in the human environment (see Figure 2.3). Agriculture is only about ten thousand years old. The Industrial Revolution started only about two hundred years ago.[39] And we can easily reel off a list of innovations that have even more recently transformed daily life, including the telephone, the

FIGURE 2.3. Technological Innovation Is Redefining Our Psychological Environment at a Pace Much Faster Than Natural Selection.

Photo credit: Tammy McGary

automobile, and the computer. Given that natural selection is a slow process, it would be miraculous if all of our psychological adaptations were well suited to our postindustrial life.

It is easy to yearn for simpler hunter-gatherer days. Yet understanding the contemporary depression epidemic requires that we travel farther back into evolutionary time and consider where the capacity for depression comes from. This is challenging, not only because we lack a time machine, but also because psychological adaptations like mood do not leave fossil remains. The best way to see that depression has deep evolutionary roots is to examine the evidence for commonalities in mood across the animal kingdom. We will consider evidence that other species can become depressed. In doing so, we have to overcome the (foolish) historical tendency to object to the existence of emotions in other species.

CHAPTER 3

What Other Species Tell Us About Depression

Six days after they gave away his sister, Ollie still won't eat his chow. He isn't even interested in a treat. When his owner shakes a sock at him in a bid for tug of war, his favorite game, he stares blankly. The doorbell rings. Ollie doesn't bark or bother to investigate.

Is this depression?

Depression in animals has long been a hard sell. In the wake of René Descartes, an enormous gulf opened between humans and other species, and Cartesian thinkers ever since have argued that other animals are mere automata, furry robots. Skepticism about complex inner states in other species has endured even into the twenty-first century. The torch has been passed from behaviorists, who wanted to banish all notions of motivation from scientific purview, to contemporary neuroscientists, who accepted basic motivational drives but not anything as elusive as animal feelings, and finally

FIGURE 3.1. The Scientific Community Has Been Strangely Slow to Open Itself to the Possibility That Other Species Experience Low Mood States.

Photo credit: AZAdam

to cultural psychologists, who have no place for animal depression, but for different reasons. For them, depression is a shared understanding, a historical artifact defined by human words and deeds.

Mood science seeks to refute these views. Ollie *is* depressed, and we know why he lost his mojo. The loss of a regular playmate is a major shock to social resources (especially for a highly social animal); it signals the possibility of other losses and uncertainty about the future. It is best to hunker down and wait, at least for a time.

Our fellow mammals, be they rats, cats, or bats, provide the most compelling and dramatic evidence for depression in the animal kingdom (see Figure 3.1). High and low moods equip these animals to track opportunities and resources in their environments; the capacity for mood is essential for guiding behavior in a changing world. This capacity also exists in other vertebrate species—the idea that birds, frogs, and fish become depressed can readily be defended.[1] Invertebrates (animals that lack a backbone) have simpler and less flexible behavioral repertoires. At best, they possess general tropisms that are precursors to mood; for example, an amoeba can move toward a nutrient gradient. For these reasons we won't be discussing depression in Ollie's fleas.

Mammalian depression runs the full gamut, from relatively brief and mild shallow depressions to severe and long-lasting deep depressions. To judge where Ollie's reaction falls along this range, we would, as with a human sufferer, need to carefully assess the number, intensity, and duration of his depressive behaviors.

A good starting point is the official diagnostic manual of psychiatry and psychology for humans (now in its fifth edition, abbreviated DSM-5), which lists nine symptoms that are components in a diagnosis of depression. Most of the human criteria can easily be interpreted in terms of canine behavior:

Human Symptom	Signs in Ollie
Depressed mood	Drooped posture
Loss of interest or pleasure	Reduced play, less interest in food and sex
Sleep disturbance	Less sleep, restlessness at night
Guilt	—
Low energy	Less vigorous on a walk; won't fetch
Psychomotor changes	Slower movement
Inability to concentrate	Lack of attention; won't perform old tricks
Change in weight or appetite	Eating less, losing weight
Thoughts of death, self-harm	—

The only symptoms on this list that are problematic to assess in Ollie are excessive guilt and suicidality, which depend critically on language. People who deny depression in other species naturally focus on the language-based symptoms like guilt and the obvious difficulty of assessing them in a household pet. The idea that your cat could express regret at being a lousy mother to her kittens might seem laughable, but even here scientists debate whether some aspects of regret might be possible in our fellow mammals.[2]

As I discussed in chapter 1, low moods exist apart from the language used to capture them. Kindergarteners, for example, lack a nuanced conception of guilt and mood and typically have difficulty reporting on these internal states in themselves. Yet it is a tragic fact that six-year-old girls and boys can be seriously depressed, and there is even increasing recognition of bona fide depression in 1–2 percent of preschool-age children.[3] It is also worth noting that people who live in diverse cultural settings often have extraordinarily divergent understandings of concepts like guilt or mood. Does it matter that in Tahiti there is no word for guilt?[4] Although the local expression of the affective (emotional) disorder may vary from place to place, documented cases of depression, with its attendant behaviors and bodily changes, have been observed in Tahiti and in every country ever studied.[5] Finally, the DSM-5 does not insist on the presence of the language-based symptoms. None are required for a diagnosis of a "major depressive episode."

Beyond the official symptoms of human depression, dogs and cats manifest numerous unofficial signs that are characteristic of depressed humans. Those who live with them know that reduced exploratory behavior, long hours hiding under the bed, and reduced interest in self-care and personal hygiene, reflected in less grooming or use of a litter box, are all signs that something is amiss.[6]

The parallels to human depression go beneath the skin as well. Hormonal changes characteristic of human depression, including increased secretion of steroid hormones[7] and decreased activity in parts of the immune system, are apparent in biological samples drawn from cats or dogs showing depressive behaviors. Upon examination of these animals' twenty-four-hour biorhythms, or *circadian rhythms*, we find the same sorts of changes in the rhythm of daily body temperature and in the sleep-wake cycle as we see among depressed people.[8] Although there are relatively few neuroimaging studies of cats and dogs, existing evidence shows strong parallels in brainwave patterns as measured by an electroencephalogram.[9] Parallels should be expected; mammals such as dogs and cats share our vertebrate brain organization and are governed by the same neurotransmitter systems. I challenge those who deny the existence of animal depression to find a single robust biological correlate of human depression that is absent from other mammalian species.

Pet owners' stories poignantly relate the frustrations of trying to understand and get help for a favorite companion that appears dejected; lacks energy; and won't eat, drink, or play. On Internet discussion boards we can find hundreds and hundreds of postings by desperate pet owners:

> *I'm sure my dog is depressed, as she has lost her bling!!!! She has not been the same dog since last summer when we went away one weekend & left her at home with our daughter. Our daughter went out for a couple of hours & left the dog*

in the yard. [T]he gate blew open, she got out, was lost for a
few hours, was run over (but not hurt). Had her checked at
a vet, everything was alright. She is still very slow moving
& is not the same dog. She does not greet us at the door any-
more & will stay in bed all day, if I didn't get her up & out
for a pee, etc.
     Posted by Marina on 2009-03-07 at 23:36:28[10]

Not least because pets can't self-diagnose, it is difficult to get an epidemiological sense of the true scope of severe pet depression. We almost certainly underestimate it, as depression is already underrecognized and undertreated in humans. Psychiatric problems in small animals are often trivialized, so it is easy for pet depression to fly under the radar. *Fortune Magazine* mocked Eli Lilly's decision to pursue FDA approval of a chewable Prozac for pets as the second dumbest moment in business of 2007, writing, "Thank God. We've been so worried since Lucky dyed his hair jet black and started listening to the Smiths."[11]

On top of this, there is the great diagnostic challenge. Depression symptoms like lethargy; weight loss; lack of interest in food, drink, and social activities; or even a tendency to sleep excessively can mimic symptoms of known animal diseases, and a vet must make a thorough medical examination to rule out an underlying health condition. Finally, there is real ambiguity about when to treat. How bad does a depressive episode have to be before intervention is necessary? In this respect, pets are like small children: they cannot self-refer to treatment or give us nuanced access to their internal lives. We must make our best guess about what is in their best interests.

Nevertheless, veterinarians have for decades made off-label prescriptions of human antidepressants to pets in distress.[12] Do they work? Hard to say. We lack the extensive clinical trial data on cat and dog depression that we have on humans. Moreover, until recently pharmaceutical companies have not been keen to undertake costly clinical trials for the smaller animal market. (The exception is Reconcile [chewable Prozac], which has been FDA-approved for treating separation anxiety in animals.)[13] Our veterinary data, a body of clinical wisdom, are necessarily impressionistic, but they suggest that when dogs, cats, and horses take human anti-depressants, the drugs are partially effective—just as they are in human patients.[14]

If the anecdotal and clinical evidence does not persuade us that other mammals exhibit the gamut of low mood states, there is much more. To start, we can consider the immense body of neuroscientific work on "animal models" of depression.

Dark Chambers: Animal Models of Depression

Animal models of depression are nothing if not diverse. This is due partially to the fantastic complexity of the brain and central nervous system and partially to the incredible specialization in contemporary neuroscience, which spans fields as diverse as genetics, functional and structural brain imaging, and cellular biology. As a result, there is no single agreed-upon animal model of depression. Neuroscientists often fret about the competition between research paradigms, seeing it as a sign that their field is in disarray.[15] Yet from another perspective, diversity is natural. If low mood is useful in different situations, we would expect that many experimental techniques could provide serviceable models of depression in animals.

It can be difficult not to flinch when reading about the experiments that fuel this research. To model depression, scientists must intentionally expose animals to adversity. These scenes can be hard to watch, but they also shed light on the nature of depression in other species and ourselves.

The least nasty (yet still objectionable) animal tests are relatively brief and center on an easy-to-measure depression-related behavior. Animal tests are used widely in the screening phase of drug development to help a pharmaceutical company decide which of thousands of compounds at its disposal has potential clinical utility as a human antidepressant. These tests, simulations of depressive behaviors, are used as a means to better link the genes or hormonal changes in an animal that may also be involved in human depression. Thousands of scientific studies have been performed with them. Clearly drug designers think other animals can become depressed.

The Tail Suspension Test, or "Tail Test" for short, is one such test.[16] Mice are suspended upside down by their tails, usually for six minutes (see Figure 3.2). The amount of time the mouse struggles and the force and direction with which it pulls are measured.

The Tail Test is based on the observation that when placed in an inescapable stressful situation, rodents will eventually develop an immobile posture after initial escape-oriented movements.[17] The Tail Test demonstrates the mood system's ability to demobilize effort, sometimes quite rapidly. The mouse has the goal of escaping from an uncomfortable (hemodynamic stress of blood rushing to the head) and highly

FIGURE 3.2. Three Mice Being Tested in the Tail Suspension Test.

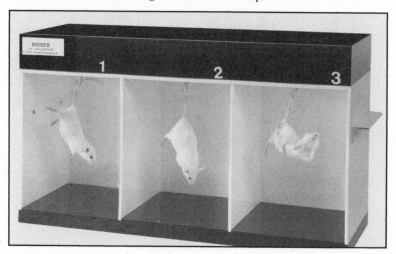

Photo courtesy of BioSeb Instruments

unfamiliar situation of being hung by the tail. Yet it learns over time that its movements cannot effect escape. Its final immobile posture is the product of a mood system that is rapidly reducing effort in the context of an impossible goal. I can assure you that if you were hung upside down from the ceiling, your mood system would reach the same conclusion. Continuing to violently struggle while in this posture would hasten exhaustion, risk of blacking out, and ultimately death.

Another widely used animal test is the forced swim test, or Porsolt Test.[18] Mice or rats are repeatedly dropped into a cylinder filled with water. This is another experimentally created situation in which escape is not possible. Again, the animal's first response is to try to escape. Upon being dropped into the cylinder for the first time, the mouse will swim vigorously, then gradually stop struggling and float with its nose breaking the surface, making minor movements only to keep

FIGURE 3.3. A Rat Floating at the Surface in the Porsolt Test.

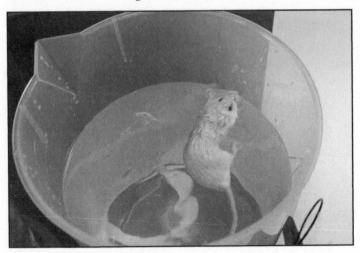

Photo courtesy of BioSeb Instruments

its nose above the water (see Figure 3.3). Upon repeated im-
mersion, the mouse becomes immobile much more quickly
and for a greater proportion of the test period. As with the
Tail Test, the simplest interpretation of the animal's immobility is that the mood system rapidly down regulates effort
to conserve energy in the face of an unreachable goal. In the
Porsolt Test, ceasing to struggle seems extraordinarily wise.
Whereas floating is possible, continued struggle could lead to
hastier exhaustion and drowning.

    Aside from showing the animals' reduction of effort,
what do these tests really model? The test developers, steeped
in the biomedical tradition, see the results as applicable to
deep depression, which they assume is a breakdown in functioning, namely a disease. Drugs that prolong escape-related
behaviors during these tests are considered new candidates
for treating humans who have clinical depression.[19] Indeed,

we can expect that when a rodent is pretreated with one of the known human antidepressants, it will almost invariably prolong the time it persists in escape-directed behaviors compared to animals treated with a placebo.

Of course the logic that "if an antidepressant changes a behavior, it must have been clinical depression" is far from airtight. Antidepressants affect many behaviors that are unrelated to depression; these drugs are used in humans to treat chronic pain, obsessive compulsive disorders, and eating disorders, among other conditions, and have even been shown to influence the behaviors of healthy nondepressed volunteers who take them.[20]

In the end, to grapple with what these tests model we must raise a broader philosophical question: Is struggle always a sign of health? Culturally we have been taught that the prolonged struggle of a mouse or a man in an impossible situation is noble. Sisyphus pushes his rock up the hill only to have it roll back down. Lacking the data to prove that persistent behaviors are always adaptive, I lean toward the opposite view—that the rodents' mood systems' immobility response is probably healthier than prolonged struggle. It certainly helps them survive longer under these conditions. If so, antidepressants are essentially overriding an adaptive response, not reversing a disease state. Clinically there is an unresolved tension: How do we weigh the cost of suppressing a natural instinct against antidepressant medications' benefits in reducing symptoms?

These tests, as much as they tell us about the mood system, cannot serve as full models of deep depression because of their brevity. The critical stressor phases last fifteen minutes at most. In common parlance, people might say they are *"so depressed"* when they are having a really bad day. In the real world, when we speak of a clinically significant deep depression, the moods are more durable: they take weeks to root and often much longer to uproot. What these rodent tests appear to induce is shallow depression. Consistent with this interpretation, the effects on animals' behavior and physiology dissipate quickly after the stressors end.

Interestingly, this matches what we see in humans when a state of mild depression is experimentally induced (albeit via less morally objectionable means). For example, when people are made to fail at a laboratory task, like having to complete anagrams that are impossible to solve, we can expect that most study participants will report feeling bad, show downcast faces, and perhaps give up on the task completely after a few minutes of flailing about. In the moment, these low mood states are the real deal, full of the sick and sinking feeling of failure. Yet these states rapidly dissipate; the participant leaves the study no worse for wear, and we have no reason to think these procedures have any lasting effects (i.e., no nightmares about anagrams).

The human and the rodent data tell us that, at least under ordinary circumstances, when a challenge ends, the mood system, ever forward looking, tends to right itself. The fleeting nature of moods can be striking. Think about watching the news on TV—it can be disconcerting how quickly a brutal, wrenching, and seemingly unforgettable image, such as a starving child covered in flies, can fade from consciousness once the TV remote is clicked off. But this raises an obvious question: If the mood system is so resilient, how do animals (and humans) fall into deeper depression?

Deep Depression in Other Species

Seeking more faithful and realistic models of clinical depression in animals, neuroscientists have developed paradigms to stimulate longer and more stable depression in their test subjects. This work naturally takes us to darker chambers, where animals are subjected to stressors that are unpredictable, severe, and/or long lasting.

One milestone in this work was the research conducted by psychologist Martin Seligman. He and his colleagues designed studies of "learned helplessness" in a variety of animal species in the 1960s and 1970s.[21] Many of their test subjects were dogs, who were exposed to shocks; in a typical experiment, they were given sixty-four shocks that lasted five seconds each. The critical element was that the shocks were inescapable. Most of the shocked dogs displayed symptoms we would recognize from human depression, including being less hungry, losing weight, moving less, and sleeping less. Furthermore, exposure to the inescapable shock diminished the animals' ability—tested in the next phase of the study—to learn how to get away from an escapable shock. Inescapably shocked animals were termed "helpless," enduring shocks passively, waiting for them to end, even when it was now possible for them to escape. These behaviors seemed analogous to human patients with clinically significant depression, who, as we will see, become so pessimistic about effecting change in their environment that they will often cling to the belief that there is no use trying. Importantly (and unlike in the Tail and Porsolt tests), the effects of exposure to uncontrollable shock persisted well beyond the experimental situation, even up to two or three days later.[22]

Yet two or three days is still briefer than the timescale of deep depression, which lasts weeks, months, and sometimes years. Intuitively, one might expect to see longer-lasting depression when unpredictable adversity continues. Likewise, one bout of inescapable or unpredictable stress should take a lesser toll on mood than multiple bouts over time. This is the difference between having a boss snap at you on one occasion and coping daily with a boss who presents an ever-expanding list of unreasonable demands, then coming home to care for a child who has developed a mysterious health condition, while fending off harassing telephone calls from creditors. Not surprisingly, a procession of unfortunate events has been shown to predict human depression more strongly than a single unfortunate event.[23]

Consistent with our intuitions, this pattern is true of other mammals as well. Paul Willner and colleagues have examined the effects of extended unpredictable stress on rats by developing a punishing routine that lasts many weeks. Rats taking part in these "chronic mild stress" experiments endure periods of food and water deprivation, multiple nights in which the lights are not turned off, cages tilted at a 45 degree angle, being paired with unfamiliar animals, living in cages in which the sawdust bedding is wetted down, being blasted with intermittent loud noise, periods in which an annoying strobe light is flashed, exposure to the cold, and reversal of the light/dark cycle. Rats that undergo weeks of this regimen secrete more stress hormones and are less reactive to an acute stress such as a loud sound. When offered a sweet drink that rats ordinarily fancy, they drink far less of it than nonstressed rats. These rats also have a reduced drive to seek pleasure, a key symptom of human depression.[24] Interestingly, it is the variety of the stressors in the chronic stress routine that really matters. When subjected to simpler versions of the routine, with just one or two stressors, the rats habituate, or adapt, to the stress.

One reason that the chronic mild stress routine is a strong animal model of depression is that once rats are established in the routine, their pleasure seeking is diminished for many months. However, if the stressed rats are administered antidepressants for several weeks, their responsiveness to a variety of rewards returns.[25] The slow pace of improvement matches what is seen in the clinic among depressed humans, who typically take several weeks to respond to drug therapy. As the rats improve, they drink more of the sugar solution than before, prefer to go to places where they have been rewarded in the past (a phenomenon called *place conditioning*), and work harder to give themselves a direct electrical brain stimulation in specific areas of the brain that scientists strongly believe are involved in pleasure.

Interestingly, not every animal that undergoes the chronic stress regimen will show prolonged signs of depression. Yet that does not mean there is a major flaw in the test paradigm; there is similar variability of response in other animal models of depression. In fact, one-third of Seligman's dogs learned normally after exposure to the inescapable shock. Mammalian depression is not a *reflex*. Therefore some variation is to be expected.[26] Situations that are strong enough to produce long-lasting behavioral deficits in nearly all animals are unusual, in the lab and in life.[27] One study of exhaustion stress in rats that produced signs of depression in nearly all of the subjects employed a stress so horrific that it killed half of the animals right away.[28]

That animals vary in their responses to laboratory-imposed adversity reveals an important feature of the mood system, one we can apply to humans. When we think of situations that strongly drive the human mood system, such as facing a life-threatening illness, dissolving a long-term marriage, or enduring a major public humiliation, they all leave room for individual differences in reaction. A full understanding of the mood system must take into account these biological, cultural, and social factors that explain how easy or how difficult it is for a given person to transition in and out of low mood, a set of issues I take up in the chapters to come.

Low Mood and the Rending of Social Bonds

Most efforts to model depression in other species test animals in solitude. This is for purely practical reasons: single-subject experiments are simpler to run, easier to control, and more readily interpreted than experiments that allow subjects to interact with one another. More surprisingly, most human experiments in induced mood also involve testing subjects individually. Yet we led off with Ollie's reaction to losing his sister, and with good reason. Complicated as they may be to study, social situations are the strongest drivers of mood.

Perhaps the most devastating observations of depression in other species involve the effects of social separation. Harry Harlow's controversial studies from the 1960s on baby rhesus monkeys raised for six months in total social isolation are difficult to read. They describe animals that appear profoundly depressed during the isolation period and are unable to function normally in group life when returned to the company of other monkeys.[29] Repeated social defeat, or physical isolation, is another potent driver of low mood. Experiments show that if an intruder rodent is aggressively and repeatedly attacked by a home-caged animal defending its territory, the attacked intruder will show significant signs of depression.[30]

Some of the most arresting observations of separation come from the wild, where the impact of social bonds on mood can be observed in a natural setting. Baby chimps and other monkeys have strong, reliable sequential reactions to being separated from their mothers. For the first day or two the baby displays signs of protest: a period of agitation, screaming, distress calls, and an inability to sleep. This is followed by a period of despair in which the baby monkey shows reduced activity; has a hunched, even collapsed posture; and is less likely to play, eat, or even vocalize.[31] Separation experiments with adolescent monkeys who have been housed together suggest these stages are not special to infants: when adolescent housemates are separated, they show the same protest-despair sequence.[32] This separation sequence is common among mammals and is seen in cats and dogs, rodents, squirrels, and human babies.[33]

Just as Harlow studied the effects of separation on monkeys, John Bowlby, a child psychiatrist, made similar observations of human infants.[34] Bowlby's primary interest was in how children form attachments, and he studied this, in part, by examining situations in which attachments were strained or broken. Bowlby, keenly sensitive to child suffering, went to orphanages that had taken in babies recently separated from their parents. His descriptions of their behaviors, captured in his multivolume classic, *Attachment and Loss*, demonstrate a remarkable continuity with the sequence seen in other primates facing separation. First is a protest phase of crying and thrashing about, then a "despair" phase, in which the child is still preoccupied with his or her missing mother, but his or her behavior suggests increasing hopelessness. As Bowlby described the despair phase, "He is withdrawn and inactive, makes no demands on people in the environment, and appears to be in a state of deep mourning. This is a quiet stage, and sometimes, clearly erroneously, is presumed to indicate a diminution of distress."[35]

The mood system that we share with our fellow mammals remains exquisitely sensitive to any social loss that may imperil survival or threaten our life plan. Whether we are helpless infants or old married couples, we are wired to detect threatening situations. The prospect of irretrievable social loss provokes low mood and makes us stop, reassess, and get help, if needed, to make a possible change of course.

With the recent spike in serious depression and stories about suicide filling the headlines, it is tempting to see depression as a modern scourge. Yet human depression derives from something very old; it is an elaboration of a primal response to adversity that we share with our mammalian cousins. With this in mind, I turn next to the most powerful cross-species trigger for low mood, the death of a significant other.

CHAPTER 4

The Bell Tolls: Death as a Universal Trigger

WHEN WE THINK OF BEREAVEMENT, OFTEN WHAT COMES TO MIND first are the cultural trappings of death. Coffins. Garden cemeteries. The repeated condolences: "I'm so sorry." The color black. Funeral parlors. The raucous storytelling at an Irish wake, or the Jewish custom of covering mirrors and sitting shiva. No matter the tradition, the experience of low mood and depression are a part of losing someone. To think clearly about depression, we need to dwell on mood changes caused by loss. To see why, we turn to bereavement in its rawest and most essential form—in animals.

Take a chimpanzee mother whose baby has died.[1] Captured on video by a team of scientists in Guinea, she stands over the body. She tends to it. She uses a leafy branch to bat away flies that are buzzing around. The film clip is short, but the scientists who shot it verified that the mother stayed with the body for several days. The end of a significant bond is a universally important event.

Mother gorillas living in captivity have been witnessed doing similar things in the wake of a child's death.[2] Gana, an eleven-year-old gorilla in the Munster Zoo, was observed carrying her dead baby, Claudio, around with her, sometimes in her arms, sometimes on her back. She would prod and stroke the body as if hoping for life to return. It did not.

Does a gorilla grieve? All signs point to yes—a gorilla that doesn't eat, doesn't sleep, doesn't explore, and seems focused on what has just been lost certainly exhibits symptoms akin to those of human grief. If we took biological samples from this mother, we would undoubtedly see the same elevations in steroid hormones apparent in grieving humans.[3] Across mammalian species, separation and loss drive telltale spikes in the markers of stress.[4] Taken together, the behavioral and chemical indications suggest that gorillas can experience low mood in the wake of a child's death.[5]

Yet why, from an evolutionary perspective, would it be that "grief and sadness," as French writer Alphonse de Lamartine observed, "knits two hearts in closer bonds than happiness ever can?"[6] Most mammalian species are sociable creatures that form strong and lasting affectional bonds, to a mother, to a pack member, to a mate. Dissolution, or threatened dissolution, of a core bond provokes immediate distress. If the bond cannot be restored—as in the case of death—low mood sets in. As Queen Elizabeth II once said, "Grief is the price we pay for love."[7]

The mood system monitors key relationships because in social species, relationships are essential. Others of our kind, our conspecifics, make key contributions to survival and reproduction such as finding and providing food, vigilance and protection against danger, and rearing of the young. A death leaves the living with *reduced fitness*, a shorthand phrase that means fewer resources for survival and reproduction are now available. After the death of a group member, the living must come together to rebuild these resources.

The loss of a baby is a particularly devastating event for a mother, an evolutionary analysis would predict, because it is a major blow to fitness and directly lowers the chance that she will pass on her genes. Data gathered from several human cultures confirm that the most severe grief is brought on by the death of a child. And the death of a child nearing reproductive age elicits the strongest response of all,[8] another clue that grief is related to fitness.

Loss of a parent or provider also reduces fitness. Surviving kin may have trouble securing food or warding off predators. Unless the group can recruit help, its survival is imperiled. Low mood is a strong signal that survival may be at risk. Even the death of strangers can cue low mood. In the ancient past, when one of your species died, it usually meant the environment was hostile to survival and reproduction: there may have been diseases or enemies about. At such moments it was wisest to hunker down, at least for a time.

Among humans, death is a powerful driver of low mood (see Figure 4.1). One could argue that every culture develops elaborate rituals around death to channel, control, and contain low mood. Because we can put thoughts and emotions into words, the death of a significant other (including nonromantic others) can be an even more potent mood driver than in other species. Humans can intentionally remember a person who was recently lost, bringing his or

FIGURE 4.1. Grief Is the Prototypical Driver of Low Mood.

US Navy photo by Mass Communication Specialist 1st Class Leah Stiles

her history to mind. Language gives us the tools to ponder the meaning and implications of a permanent loss. Indeed, thoughts of the deceased and life without him or her often bubble up, unbidden.

Just about everyone who mourns a significant other will experience low mood. The duration varies—from hours, days, and weeks, to months and sometimes years. Although most people who grieve the death of a significant other escape having a disabling depression, nearly one in three will fall into a clinically significant episode.[9] Given that none of us can avoid exposure to death, it will invariably be the trigger for many depressions. Even with imperfect statistics, we can estimate that nearly one-fourth of all depressions are bereavement related.[10]

Given these numbers and bereavement's near universality, it is not difficult to make a strong case for studying human responses to death as a means to understand low mood and depression. Indeed, if the prototypical situation that generates low mood and depression is an irrevocable loss, and if the prototypical loss event is the death of a significant other, then logically, bereavement would be a central topic in depression research. Yet it is marginal. For example, in 2010, of the 404 articles published in the *Journal of Affective Disorders*, a leading psychiatry journal that covers all aspects of mood disorders, only three considered bereavement. Similarly, the recently published *International Encyclopedia of Depression* devotes a mere three pages (a single entry) to bereavement, out of the tome's 574 pages. These figures are representative of bereavement's status when it comes to the study of depression. The two are rarely mentioned in the same breath.

This is so because until very recently, bereavement and depression have been in entirely separate worlds when it comes to research. The people who study depression and those who study bereavement are by and large in different fields, publish in different journals, and attend different conferences, and they are ultimately animated by different research questions and concerns.

Striking evidence of the gap between depression and bereavement is the historical awkwardness with which our modern diagnostic system handles low mood in the context of a death. For example, for decades, in the official bible for mental disorder diagnosis, the DSM, depression within two months after a death was usually not termed "depression." Instead, depression following a death was considered under another category, called "simple bereavement," which was not indicative of a mental disorder or condition.[11] In fact, of all the things that can befall a person, bereavement has historically been the only life event that could potentially negate a diagnosis of depression.

What's going on? To understand, we must return to the first premise of the diagnostic book: mental disorders reflect diseases and are not part of normal variation. By the tenets of the disease model, the symptoms of depression must reflect abnormal functioning. The problem is that the architects of the DSM system were well aware that bereavement is a situation in which it might be typical, and possibly even adaptive, to experience significant depression, at least for a time.

Bereavement-related depression creates a quandary for the disease model, because bereavement produces the same symptoms as the putative disease without reflecting any abnormality in functioning. Thus, excluding bereavement-related depression (presumed adaptive) from the category of depression (presumed a disease) was important for maintaining the premise that all DSM-cataloged mental disorders are dysfunctions.

It is a weighty enterprise to draw bright lines between depressions that are meaningful and even healthy, and others that are not, when the person's symptoms are considered products of a dysfunction, the "electrochemical static" of the disease model.[12] Marking off bereavement depressions as separate from other depressions is not an esoteric diagnostic exercise; it affects how millions of people are labeled, treated, and judged. We would expect such a division to be based on extensive evidence that bereavement and nonbereavement depression differ in important respects. Yet this is not the case. Only recently have scientists reexamined the bereavement exception in light of real data that compare people who are depressed because of bereavement with people who are depressed for other reasons. If such tests find that depression is similar across different trigger events, that would support the idea that our mood system is configured to respond in broadly similar ways to any major loss—be it of a significant other, a job, a relationship, or reputation.

What we are learning does indeed support this notion: whether it is prompted by the loss of a spouse or of one's life savings, depression is depression is depression.[13] Informed by these new research findings, in 2013 the mood disorders committee at long last finally eliminated the bereavement exception in the newest edition of the DSM (DSM-5). No longer blinded by this exception, we can see depression in a more unified fashion and understand the key role bereavement plays in it.[14] Let's review what led to this reversal.

Depression Is Depression After All

In a first strike against the validity of the bereavement exception, psychiatrists Sidney Zisook and Kenneth Kendler wrote a comprehensive review article in 2007 about bereavement-related depression.[15] Because few studies at that time had directly compared bereaved-depressed and nonbereaved-depressed samples, the authors used an indirect strategy, simply comparing what was found in separate studies of the bereaved depressed to our existing knowledge base about regular depression.

In opposition to the some-depressions-are-diseases view, bereavement-related depression had observed characteristics that were strikingly similar to regular nonbereavement depression. Many studies revealed that people who suffer from bereavement-related depression receive less support from others, just as we see in regular depression.[16] They also tend to be in poorer overall health, much like people who suffer from regular depression.[17] When objective measurements are made of the body, bereavement depressions are associated with many of the same bodily changes that we see in regular depression, such as dysregulation of steroid hormones, immune changes, and even changes in the electrical activity of the brain during sleep.[18] Finally, there is even similarity between the clinical courses of bereavement-related and regular depression. Bereavement-related depressions last for about as long as regular depressions do. People who have a bereavement-related depression are about as likely as those who have a regular depression to have additional episodes of depression in the future.[19] And though there is great controversy about treating bereavement-related depressions, when interventions have been tried, these depressions have been about as responsive to interventions (such as drugs or interpersonal psychotherapy) as are regular depressions.[20]

In the years since Zisook and Kendler completed their review, data from better-designed studies have rolled in, including from studies that perform the direct key comparison between persons who have bereavement-related depression and persons who have depression from other causes. Jerome Wakefield and his colleagues published a major article in *The Archives of General Psychiatry* based on a representative sample of the United States that contained 8,098 persons aged fifteen to fifty-four years. The authors reexamined this vast data set, which included extensive psychiatric interviews of all participants. They were able to identify persons whose depression was triggered by a death and contrast them with persons whose depression was triggered by other losses, such as the loss of a job or the end of a marriage. In these national US data, depression triggered by bereavement and depression triggered by other losses were again astonishingly similar: the two depressions had similar symptom profiles and were comparable in terms of how long the symptoms lasted, whether or not the person attempted suicide, and whether the person sought out mental health services for his or her difficulties.[21]

Similar findings were independently reported by another group, using a vast sample of twins in Virginia.[22] Bereavement-related depression was similar to nonbereavement-related depression across a host of measures; the twins did not differ on age at onset, on how many prior episodes of depression had been experienced, on how long the symptoms lasted, on their risk for future depression episodes, and in what other psychological problems could be diagnosed. The two groups were even similar in their personality patterns: those who suffered from bereavement-related depression were just as extroverted as those who suffered from depression from other causes. In an added wrinkle, the authors used the twin design to examine whether bereavement versus nonbereavement in one twin would differentially predict how much depression the other twin experienced. The disease model strongly expects that if one twin has regular depression, this should predict depression in his or her co-twin much more powerfully than if the first twin had depression because of a death. Depression after a death should be less predictive, because these bereaved twins presumably do not have the depression "disease." In another blow against the disease model, this was not the case. Whether the first twin experienced bereavement or non-bereavement depression had no bearing on *how much* depression the co-twin experienced.

Finally, these findings are from the United States, and you might wonder if things are different elsewhere. Given cross-cultural differences in how people grieve, bereavement-related depression may take different forms in different cultures. Nevertheless, these basic findings have been reproduced in several countries, including Lebanon, Denmark, and France. As yet no evidence has emerged to suggest that bereavement-related depressions are substantially different from other depressions.[23]

The Road to Understanding Depression Runs Through Bereavement

Free of the misconception spread by the bereavement exception, we can see that the links between bereavement and depression are fundamental and are so strong that one might even say the road to understanding depression runs through bereavement. Bereavement is the most universal and most potent trigger of low mood; it commonly leads to clinical depressions that are otherwise indistinguishable from depressions that arise from other triggers. In fact, if we step back, we can see several ways that bereavement provides a model for thinking about depression, offering clues to what drives the mood system and helping us predict when bouts of intense low mood will occur.

Providing Clues to Triggering Events

It was a summer day in Nevada. "He said there were so many pine needles up on the roof that he was going to get the blower and get up there. I said, 'Please don't go up there because you are going to fall.' He had a bad hip, and he had had the other hip replaced. But he said no, it would only take a few seconds." A moment later, Kenny Guinn, the former governor of Nevada, would fall off the backyard roof of his Las Vegas home, widowing Dema Guinn after 54 years of marriage, and plunging her into an unrelenting depression. Still suffering and reluctant to leave the house nearly five months after that tragic day, Dema asks, "Dear Lord, why take Kenny in such a senseless accident?"[24]

Most depressions grow out of events in the world around us. Although few are as horrifying as witnessing the violent death of a loved one, almost nine out of ten depressed people can identify external events that in some way contributed to their depression, with more than half reporting a severe, stressful life event prior to the onset of depression.[25] Negative events cannot provide a full explanation for why people become depressed, but as we will see, for most sufferers they play a role in setting the mood system on a path toward depression.

Bad life events come in great variety and have many themes: uncertainty, danger, humiliation, injustice, and so forth. However, when people's reported life events are rated on different themes by objective coders, the theme that most consistently predicts depression is loss. You can lose your livelihood, reputation, or marriage. But the ultimate loss, bereavement, is the prototypical loss event, the one that most strongly predicts depression.[26]

The idea of a "loss event" is posed in the singular, but, if we scratch the surface of any event that ignites serious depression, there are often multiple losses simultaneously driving the mood system. For Stacy Murette, who lives in a small Wisconsin town, the loss of her long-term marriage to divorce was a cluster bomb of loss. "When Bob left, everything fell apart," she says. The end of her marriage meant first and foremost the loss of a special relationship. After nineteen years, "he didn't love me and he wanted a divorce." It also represented the loss of Stacy's dream: when she got married after her freshman year of college, her "life held a lot of promise" because in her mind, "the goal was the family." The divorce was cataclysmic, shattering her ideal of family and that of her two boys and two girls, who "realized that Daddy just isn't the father they want and more importantly, need." Piled on top of these losses were blows to her imagined future, her strong religious faith (divorce runs counter to her religious beliefs), and her economic security. It was little wonder that when Bob left, Stacy plunged into a depression that raged on for four years.

Although some cases of depression lack obvious loss events, the theme of loss is often still present in more subtle ways. Take a young adult's depression that emerges after he starts working at an ordinary job after college; the depression might be related to the fact that taking a less-than-ideal job meant giving up on a childhood career dream, even if the loss of the dream was not discussed explicitly and the young adult is only dimly aware of the connection between the loss and the symptoms. As discussed in Chapter 2, the mood system is open to a large number of simultaneous inputs, and because of this we can only make probabilistic statements about why certain moods prevail. This can obscure the significance of a loss.

Clinicians in the field have noted the therapeutic importance of "uncovering" an unaddressed loss event in a client. John Grace, a psychiatrist practicing in Florida, said, "I have had a number of patients I've done CBT [cognitive behavioral therapy] for months on and finally after spinning my head in circles I realize I have been missing a tremendous loss for them. For example a woman who moved to Florida and is missing her job and her loss of esteem. If she comes in she might gloss over this loss, citing problems with marriage or energy. Until we talk about it and, in effect, do grief therapy, she stagnates."[27]

Shedding Light on Typical
Depressive Behaviors and Their Functions

The core of what people do when they grieve a death is to focus on the loss and its significance. During bereavement this focus may be so strong that the griever is flooded with thoughts, images, and memories of the dead person, which sometimes extend to feeling his or her constant presence nearby or at a distance. As one mourner described it, "I know that when I look up in the sky at night and I see a bright star shining I will know that it is my mom giving me the strength to go on."[28] The bereaved may search for the just-departed and feel the urge (sometimes distressing) to call out for him or her.

The flip side of intense focus on a loss is diminished interest in virtually everything else. Whether it's television, news, or sex, the mourner may have difficulty focusing on other activities or topics. As grief strengthened in the winter after her father's death, Tracy Thompson described her waning zest for living: "As the months went by, the breathtaking reality of my father's death became a physical hurt, a heaviness in my bones, a pervasive lethargy. I slept long, long hours; when I was awake I comforted myself with food. . . . It was, though I did not know it, the blanketing of depression."[29]

The same might be said for those who experience depression after a romantic breakup. The rejected party is troubled by thoughts and images of the former lover, obsessively replays conversations, and wonders whether the breakup was inevitable. Indeed, when we think of funerals, gravestones, cemeteries, and all the rituals involved in grieving a death, it's easy to overlook the fact that similar behaviors accompany other losses. For example, those who suddenly lose their fortunes (an all-too-common occurrence these days) pine for the lost lucre and engage in counterfactuals about what might have been ("if only I had sold earlier"). Across the wide range of loss events, depressions involve grieving behaviors (see Figure 4.2).

Providing Clues to Stopping Depression

Although nearly everyone experiences low mood right after a death, only 10–15 percent of people grieving death-related loss are still depressed one year later.[30] What can be learned from the fact that bereavement often does *not* become depression?

An important lesson is that the mood system typically has natural resilience. As Sophocles wrote, "Gentle time will

FIGURE 4.2. Humans Mourn a Wide Variety of Losses.

Photo credit: David Bleasdale

heal our sorrows." Healing in grief is likely sped up by culturally based rituals and rites that provide social and acceptable outlets for grief to take place. Sadly, these outlets are not as available to people laid low for other reasons. Bereavement-related sorrow leads people to deliver casseroles. This is less often than the case when a sufferer is laid low for other reasons. I wonder how many depressions might be arrested if there were a greater number of acceptable outlets for sharing it.

Yet there are reasons for hope. Activities that people engage in to resolve grief offer clues to how to stop low mood from escalating in other situations. As we consider the tenacity of depression, two features of bereavement point to the road ahead. First, the resolution of grief often involves changes in cognition, an active process in which one re-works the meanings of events. A difficult task in grief is to forge an acceptable meaning for a death, especially when it is untimely, painful, or violent. This speaks to the challenge seen for other triggers of low mood, which similarly force an assimilation of new, and often painful, information (such as the reality of rejection, failure, or humiliation). Second, resolving grief often involves filling behavioral gaps. A person has departed; routines are disrupted. The bed is empty. There is less dinner conversation. Often the key to resolving the loneliness and longing of grief is venturing forth to seek out new people or renewing old relationships that eventually fill these gaps. So, too, in other depressions, behavioral changes often provide levers that help people find a way out. With this in mind, we'll explore the behavioral gaps opened up by breakups, job losses, and other social rejections; the ways that people fill such gaps; and to what extent filling them can arrest low mood before it deepens into full-scale depression.[31]

Providing Clues to How Depression Starts

Finally, as we consider the rising arc of a depressive episode, what we have learned about bereavement also helps us think about how depression starts. Depression rarely starts from nothing. Certainly after a death larger losses that require major life adjustments will drive the mood system much harder than the loss of minor relationships. But the magnitude of the loss event is only part of the story. Depression is more likely to follow death if the seeds of depression were already planted. It is more likely to happen to people who generally have trouble coping with stress, to people who by nature have a moody or nervous temperament, and to people who have harried daily routines that don't allow for sufficient sleep. I turn now to these seeding grounds for low mood.

CHAPTER 5

The Seedbed
of Low Mood

Sylvie had already been through far worse. Her mother's cancer death when she was 12. Her husband's struggle with addiction culminating in a fatal overdose in a Delaware motel room. Her daughter, four at the time, would never really know her father. Ten years figuring out how to be a single parent and a working mom, a professional in the field of social work. Now her precious daughter, Madeline, was going through a high school rebellion. She was getting into trouble, challenging her teachers. I can't rein her in, Sylvie remembers thinking; everyone in the family was telling me, "You should punish her," and telling me I was doing a bad job of parenting. Maddie didn't want to engage; she closed me down, pushed me away. I finally said to Madeline—"It's up to you to get an education."

Three months later, Madeline was back, asking for help. Her grades were poor, and she was on the verge of flunking

math and French. Sylvie got her tutors. She started to worry,
to obsess. "What if she fails?" "It's my fault; I didn't get her
help sooner." "What if all the work I did as a mother until
the age of 14 was for nothing?"

Sylvie wakes up at 3 or 4 AM, her mind full of brood-
ing thoughts; there is no way to get back to sleep. At work
Sylvie battles fatigue; her clients notice something amiss.
They ask, "Are you OK?" She pinches herself to stay awake
in session. After several weeks of feeling like her life is out
of equilibrium, she visits her primary care doctor. He pre-
scribes Ambien.

Sylvie's case also reminds us of the ways that depressive symptoms are neither wholly good nor bad. The symptoms make her focus on an important problem, Maddie. For a mother, school failure of an only child is a keen Darwinian dilemma. Anxiety and sleep disturbance send the insistent and obvious message that Maddie's future is in peril; rest can wait. For Sylvie, being there for her daughter at a critical moment is even more poignant; she lost her own mother at nearly the same age, and now she can be the mother that her mother could not. But if Sylvie's symptoms serve a higher purpose, they also have obvious costs: her productivity at work takes a hit; she finds her insomnia distressing; and she feels "out of control," "not herself."

We have all known people like Sylvie, people with shallow depression who are struggling to carry on and are burdened by a few nagging symptoms. Yet strangely, clinical science and practice have paid little attention to their plight. There is far less research on shallow depression than on deep depression, and the causes of shallow depression have been less rigorously studied. Some clinicians even question whether low-level symptoms of depression *should* routinely be treated. Isn't this just the normal suffering that is to be expected in this universe, the collateral damage of life? Indeed, social critics like Jerome Wakefield and Allen Horwitz contend that our boundaries for diagnosing depression are already too loose, mistakenly catching people with ordinary worries and transient adjustment difficulties in the net of "true" psychiatric conditions.[1] However easy it may be to dismiss minor shallow depression as relatively benign, even "normal," it turns out to be a crucial factor in our overall understanding of depression.

Is Minor Depression Minor?

In 2001 Mark Hegel and his colleagues studied a group of patients in New Hampshire who, like Sylvie, had reported to their primary care doctors that they weren't feeling up to snuff, complaining of a low mood, vague aches and pains, insomnia, and problems concentrating. By definition, patients with a minor depression have the same kinds of symptoms as patients with a major depression, but have fewer of them. Such people are encountered frequently in primary care, more frequently than are patients with a deep major depression.[2] Despite the frequency of minor depression, the lack of clear guidelines for how to handle such patients reveals the medical profession's ambivalence about it. Indeed, the DSM has historically taken an awkward stance on minor depression as a diagnostic category, tentatively tucking it away in an appendix for conditions requiring further study or including it in a catchall category of unusual symptomatic presentations (other depressive disorders). Perhaps unsurprisingly, there are few treatments explicitly for minor depression.

For these reasons, Hegel examined "watchful waiting," which has historically been a common practice in primary care for less-affected patients. Such patients were monitored carefully by their physicians under the assumption that most would improve on their own over a number of weeks. Treatment (usually antidepressants) was not initiated unless a patient's condition deteriorated. Hegel's study was designed to systematically test the main assumption behind watchful waiting: Do most patients with minor depression get better on their own over the course of a month's time?

Much to Hegel's surprise,[3] only about one in ten enrolled patients with minor depression became well in a month's time.[4] The rest—nine of ten people suffering from low-grade depression—remained stuck in their low mood. A multisite clinical study in Pennsylvania, Texas, and California afforded another opportunity for controlled observation of patients, this time at the outset of a drug study a month before patients began taking the study medication. In this case only about 6 percent of patients spontaneously remitted in a month.[5]

If these data refute the clinical practice of waiting for minor depression to lift, the larger-scale epidemiological data are even more convincing. Yearly interviews with 10,526 community residents in three US metropolitan areas found that nearly three in four, or 72 percent of, people who had a minor depression were still bothered by one or more symptoms of depression when they were reinterviewed a year later.[6] Together, the clinical and epidemiological data suggest that, contrary to popular opinion—and medical practice—minor depression is not that easy to shake.

The persistence of depressive symptoms helps explain in part why they are so prevalent. At any given time about one-fifth of the population, 22 percent, has at least one significant symptom of depression, such as an empty mood that won't lift or pervasive feelings of guilt. The overwhelming majority are not severely affected; people who have shallow depression outnumber those with deep major depression six to one.[7] Most are like Sylvie, able to get through the day while concealing from friends and coworkers how bad they feel. In a hasty office visit with a family doctor, the depression bell may not even ring.

Perhaps for this reason, minor depression often flies under the radar. Low-level sadness seems so ordinary that it receives little special scrutiny. Existing treatments don't necessarily do a good job of managing it; little energy has gone into designing new treatments expressly for it.

In many ways the term *minor depression* is a misnomer. Individuals with minor depression may use outpatient services, miss work, and spend days in bed just as frequently as individuals with major depression do.[8] Further, we ignore minor depression at our peril. Having a minor depression quintuples the risk for developing a deep major depression.[9] People with shallow and persistent low-grade depression constitute an immense pool of first-line recruits for the growing army of the severely depressed.[10]

Does the surprising persistence of shallow depression mean that every moody hill and valley is a dangerous harbinger of a psychiatric problem? If not, how can we sort out when, and for whom, low mood is an omen of worse things to come?

How Moods Unfold

To understand when moods are and are not problematic, we need to take a fuller inventory of moods as they play out in the world. Inventorying mood is more difficult than it sounds. It requires finding groups of research subjects who are willing to be followed around and repeatedly asked, for days or weeks, "So, how do you feel now?" Fortunately the researcher does not have to physically accompany the subject as she goes to the grocer or shows up at the office; he can prompt her with an electronic pager to record moods on the spot or ask her to provide a detailed recap of the day's moods. Daily diary studies of mood do precisely this.

One representative study[11] was conducted over forty-two days. Each day 332 respondents reported on the negativity of their mood and how many stressors had affected them. All told, the ebb and flow of mood over eleven thousand days was collected. Subsequent analyses produced both predictable and surprising results. As we might expect, when participants reported stressors, that same day's mood took a hit and was more negative. Remarkably, however, one day after the stressor, mood was back to normal. In fact, day-by-day analyses showed that subjects often rebounded to better than normal. If someone had a stress-free day after a stressful day, her mood was actually *better* than it was on other stress-free days.[12]

The bottom line, confirmed by other studies using similar designs, is that stressors take only a one-day toll on mood.[13]

Most mood perturbations are short. People usually show remarkable recovery of mood, at least from garden-variety upsets. Lovers patch things up after a spat. Bosses apologize for angry outbursts. The leaky roof gets repaired.

The brevity of mood perturbations reveals at least two features of mood. First, people are active participants in their own moods. Most of us not only notice a bad mood, we take steps to change it—whether by going for a walk, calling a friend, listening to a favorite singer, or pinpointing the cause of stress. These steps, termed *mood regulation* or *mood repair*, often limit the intensity and duration of a garden-variety negative mood. Second, and perhaps less obvious, the brevity of most mood perturbations points to a natural resilience built into the very architecture of the mood system. The mood system is, by design, forward looking. It is most interested in what to do *next*. In this sense, evolution has shaped our minds not to cry over spilled milk.[14] When a stressor proves to be less than a full Darwinian crisis with ongoing implications for our own survival or reproduction, our mood system has a bias to move on. This is why even the most excruciatingly sad movie or the grimmest nightly newscast rarely arouses more than a transient low mood.

The idea that the mood system can right itself also squares with the epidemiological research; if low mood were *always* persistent, then the pool of people with depressive symptoms would grow and grow until eventually everyone was in it. Yet studies consistently show that a much smaller group—about one in five people—are in the pool of those with depressive symptoms.

If most strong moods are short, studies with dense sampling also tell us that most moods are not strong. In one,

203 healthy nurses rated[15] their mood state at work and off
the job every twenty minutes for four days, which produced over thirty-four thousand occasions on which mood
was rated. Of those times that the nurses rated their level
of anxiety, there were only seventy occasions on which the
anxiety was rated as extreme (one-fifth of 1 percent). This
pattern was not unique to anxiety. Strong anger was similarly
infrequent. Nor are nurses somehow uniquely imperturb-
able. A month-long observation of college students sampled
their moods on more than five thousand separate occasions
and found the students reported no sadness 72 percent of
the time; when sadness was reported, it was typically experienced as mild.[16]

Thus the research offers a reassuring conclusion: most
moods are short and weak. Yet there are times when milder
low moods transform into the longer, more troubling ruts
of minor depression. What explains the persistent shallow
depression that Hegel observed in New Hampshire? If shallow depression is the seedbed of deep depression, what is the
seedbed of shallow depression?

To pursue this question we have to keep in mind that
the mood system is the great integrator. To guide behavior
effectively, it must do a simultaneous readout of the state
of the world, the body's internal milieu, and the potential
payoffs and costs of various courses of action. Given this integrative function, we would expect there to be various ways
by which a person ends up in a shallow depression. And
indeed there are; to continue with our agricultural metaphor,
at least three different kinds of seeds lie on the seedbed for
low mood.

Events That Seed Low Mood

As we have seen, most people show surprising recovery from the effects of day-to-day stress. It's also clear, however, that for some stressors, the psychological fallout is more difficult to contain. Recall Matt from the first chapter. In April of his senior year of high school, a year before the onset of his depression, Matt's parents sat him and his sister, Suzy, down. They made two announcements. First, they were getting a divorce. Second, Matt's father was not only moving out of the house, he would be moving away, to Belgium, to live with the woman with whom he had been having an affair. Matt's mother had discovered the international love letters, and, just like that, final decisions were made. Reconciliation was out of the question. Suzy was a wreck, but Matt, at least initially, coped fine. When confusing thoughts of the divorce came to mind, he would push them away, bulling through the bad feelings as he always had. As time wore on, these events and their unresolved meaning returned with fuller force, increasingly looming over Matt's freshman year of college as he slowly crept toward depression.[17]

Modern psychological theories postulate that we recover more quickly from a bad event if we can readily explain it.[18] We would expect, then, that events that generate mixed feelings and/or confusing thoughts would be a powerful impetus toward persistent low mood. Denise, a young woman from Minnesota, was cheating on her husband when she found out that he had terminal colon cancer. Her thoughts and feelings pulled her violently in different directions: the excruciating guilt of cheating on a now-dying man (was the cancer divine punishment?) was counterbalanced by the frightful burden of caring for him as his health deteriorated; the sinking feeling that she surely could not abandon him was coupled with the knowledge that she had completely fallen out of love with him.[19] Events that present irresolvable dilemmas on themes that have evolutionary significance—like mate choice—are fertile seeds for low mood.

When the bad things happen also matters. Extensive research demonstrates that early life traumas, such as physical or sexual abuse, lay the groundwork for a slow creep of depression and anxiety.[20] The creep starts very early. Jan Easterbrook, a middle-aged midwesterner, dated her low-grade depression back to when she was eleven or twelve. As a preteen, Jan experienced a dawning recognition that her father was an alcoholic, although his alcoholism was not acknowledged in the family. "Feelings were not okay at my house," she remembered, "emotions were not okay to show, let alone talk about." At the same time, Jan sensed "how dysfunctional our family was, and how responsible I was." Indeed, Jan routinely took care of her younger brother, which included occasions when her father was out drinking, and her mother was with him or on her way to pick him up. When Jan was molested by a neighbor at age thirteen, she had no one to tell. If she broke the taboo against revealing something so disturbing, only bad would follow: her mother would blame her; her father would be enraged and might try to kill the molester. She kept the trauma to herself.

Jan's chronic feelings of anxiety and sadness are natural, the product of an intact mood system. In a world in which a child's primary attachment figures—parents—are emotionally unavailable and unable to help when a trusted neighbor turns into an attacker, the mood system is ever forward looking. It assumes that, if the worst has already happened, it can and will happen again. Best to be prepared. Anxious moods scanning for danger (especially in relationships) and sad moods analyzing *what* was lost and *why* serve as the last lines of defense against further ruin.

Although such armor provides the wearer protection from slings and arrows, it's also a heavy burden to haul around. Throughout her adolescence and young adulthood Jan labored under the protective weight of low mood. She constantly felt a sense of unease in her own life. Given how she was shadowed by shallow depression, her lack of support, and her traumatic experience of molestation, it is remarkable that twenty years elapsed between the molestation and her first episode of deep depression.[21]

Temperaments That Seed Low Mood

Despite their importance, stressful events do not tell us everything we need to know about the causes of low mood. Most obviously, there are substantial differences between people's mood responses, even to the same event. When workers receive layoff notices as part of a factory closing, some react equably and experience only transient emotional distress, whereas others are devastated and experience long-lasting depression and anxiety. Systematic study of catastrophic events, such as the attacks of September 11, 2001, powerfully demonstrates the importance of individual differences. A month after the attacks, about 10 percent of residents who were living in lower Manhattan on September 11 were experiencing major depression; others were affected with less severe depressive symptoms; and still others had no symptoms at all, functioning much as they had on September 10.[22]

Increasingly the study of temperament has revealed that differences in mood reactivity can be detected early in life, even in infancy.[23] Decades of research by Jerome Kagan[24] have shown that some nine-month-old infants have reasonably consistent and strong fear reactions to a variety of potentially threatening situations, such as being approached by an unfamiliar robot toy or being requested to drink an unknown substance, whereas other nine-month-olds consistently exhibit no fear in these situations. Temperamental differences, revealed so early in life, are likely to be substantially controlled by genes.[25]

However, there is no good overarching theory of *why* humans come into the world with such a wide range of temperaments. From the perspective of evolutionary selection, variations in temperament will be preserved within the gene pool if, and only if, they do not confer systematic reproductive disadvantage. We can conclude that a wide range of temperaments—from the shiest wallflower to the belle of the ball—represents viable designs.

This point can be illustrated with the story of the grasshopper and the ant. This fable is about a grasshopper who spent the warm summer months singing and an ant who had no time for fun in the sun, instead laboring to store up food for winter. Is it better to have the temperament of the grasshopper or of the ant? If you have been paying close attention, you are now alert to the potential fitness costs and benefits of virtually any trait and will correctly see this as a trick question. In times of plenty, grasshopper types will do just fine. If there's scarcity, ants might have the advantage. I prefer to be an ant and live, even at the cost of brooding worry the rest of the year. The historical conditions of our ancestral environments, sometimes bounteous and sometimes hard, have varied enough to leave plenty of room in the gene pool for both kinds of creatures.

Experiments by evolutionary biologist David Sloan Wilson also demonstrate that there is no "single best temperament." In one condition, Wilson dropped metal traps into a pond containing pumpkinseed sunfish. A subset of the fish showed boldness and interest in investigating a novel object. This was a really bad move, as they were immediately caught, and had Dr. Wilson been a real predator, it would have meant the end of their genes. Another group of fish were wary and stayed back from the traps; they were not caught. This situation favored the wary fish.

In a subsequent condition, all the fish were scooped up, brought into a new environment, and then carefully observed. Here the previously wary fish had great difficulty adapting to novelty. They were slower than their bold compatriots to begin feeding, taking five more days to start eating. In this situation the survival of the bold fish was favored.[26]

This perspective on the advantages of temperamental diversity suggest that there are good reasons why some—maybe even many—people are born with depression-prone temperaments. The most important and well-studied depression-prone personality trait is neuroticism. People who score high on neuroticism are prone to anxiety and other negative feelings (think Woody Allen) and have stronger upsets in response to stresses, be it sudden job loss or terrorist attack.[27] There is robust evidence that neurotic temperaments predispose people to experience periods of low mood and periods of more severe, long-lasting depression. A wealth of data indicates that highly neurotic people are like the fabled ant: they worry about bad things that may or may not happen in the future, and they are more vigilant about threats, even those that are distant, hidden, or subtle.[28] These capabilities are sufficiently useful that genes coding for the neurotic temperament remain within the population[29] even if they carry other costs, such as an elevated risk of stress-related conditions like ulcers or high blood pressure.[30] Like depression itself, temperaments that seed depression are neither wholly good nor wholly bad.[31]

Routines That Seed Low Mood

Finally, mood is about the mundane. Day-to-day routines—how we spend our time, how we care for our bodies and minds—continually shape our moods and can have a strong influence on whether low mood persists. Routines that build up physical and mental resources can raise mood. Other routines, woven into the fabric of modern life, are grossly misaligned with evolutionary imperatives and have the potential to seed low mood. Many of our most familiar routines seem almost perversely designed to wreak havoc on the mood system.[32]

One mundane influence on mood is daily light exposure. After all, mood evolved in the context of a rotating earth, with its recurrent twenty-four-hour cycle of light and dark phases. Our species is diurnal, and the best chance of finding sustenance and other rewards was in the light phase (think about the challenge of identifying edible berries or stalking a mammoth). Consequently, we are configured to be more alert during the day than at night.[33] Consistent with the link between light and mood, some clinically serious low mood is triggered by the seasonal change of shorter daylight hours. The onset of *seasonal affective disorder*, a subtype of mood disorder, is usually in winter.

Our newfound reliance on indoor light has effectively turned most people into cave dwellers. Artificial light is much fainter and provides fewer mood benefits than sunlight. When small devices that measure light exposure and duration were attached to adults in San Diego, one of the sunniest cities in the United States, it was discovered that the average person received only fifty-eight minutes of sunlight a day. What's more, those San Diegans who received less light exposure during their daily routines reported more symptoms of depression.[34]

Just as artificial illumination has freed us from the light-dark cycle, it has also opened the door to night shift work, which upsets the body's circadian rhythm.[35] Electricity powers evening routines that conspire against rest. About 60 percent of adults in the United States regularly watch TV in the bedroom in the hour before bedtime, stimulation that competes with and displaces time for sleep. The profusion of electronic applications—smart phones, laptops, and iPads—provides a welter of stimulating activities that occupy the mind deep into the night. Ninety-five percent of those surveyed in the 2011 Sleep in America poll said that they use some type of electronics in their bedroom at least a few nights a week within the hour before bed.[36] Ten percent of Americans report being awakened at least a few nights a week by phone calls, text messages, or e-mails. It's no wonder thirty million Americans have insomnia!

Anyone who has tried to pull an all-nighter can anticipate the results of controlled sleep deprivation studies: mood is lower after even one night of sleep deprivation. Moreover, brief experimental sleep restriction induces bodily changes that mimic some aspects of depression.[37] It's important to ponder the consequences of sleep deprivation now happening on a mass scale: more than 40 percent of Americans between the ages of thirteen and sixty-four say they rarely or never get a good night's sleep on weeknights, and a third of young adults probably have long periods of at least partial sleep deprivation on an ongoing basis.[38] Over the last century average nightly sleep duration has fallen. In 1910 Americans slept an average of approximately nine hours; that average had dropped to seven hours by 2002.[39]

Part of the answer to the riddle of low mood, then, lies in contemporary routines that increasingly feature less light, less rest, and more activities that are out of kilter with the body's natural rhythm.[40] And although we have discussed events, temperaments, and routines separately, these seeds can cross-fertilize. A man born with a depression-prone temperament gets stuck working the night shift. A new mom has screaming fights with a physically abusive husband *and* is sleep deprived from caring for colicky infant twins. Under such hothouse conditions, we can expect that depression will grow fast.

FIGURE 5.1. Old and New Views of the Typical Course of Depression.

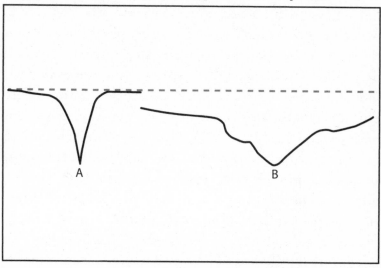

The Road Ahead

It is clear that despite the relative clinical neglect of its effects, persistent low mood, which affects millions, is the entry point into more serious depression. And this theme of persistence is vital to a greater understanding of the development of depression. Over the past two decades a flood of epidemiological data has led to a sea change in how we view the typical course of mood disorders—and persistence is the thread that runs through every phase of the course.

Figure 5.1 shows how the standard view of the course of depression has shifted. The old view (Line A) was that depression had both an abrupt onset and a fairly rapid return to a normal mood state. Mood disorders were seen as brief, time-limited crises. More recent studies with larger samples have shown that depression is more typically like Line B, which depicts a condition with an onset of low mood that is slow and creeping. In the new view, progression to a crisis of deep depression is simply a worsening of this preexisting low mood. Finally, as depression recedes, the person is not clinically well but remains mired, unable to shake off all vestiges of low mood.[41]

Informed by this new epidemiology, we now trace the downs and ups of B. Thus far, we have been on the first leg downward, the onset of low mood that so often heralds serious depression. The chapters ahead, enriched by contemporary mood science, show how the theme of persistence informs every successive phase of depression. We follow the slide into deep depression, to its lowest point, and to the phases of improvement—the painfully slow, often halting, climb out of depression.

CHAPTER 6

The Slide

To this point I have focused on environmental routes to low mood, and for good reason: onslaughts from the environment—when they threaten fitness goals—are a core cause of mammalian depression. These hits can be acute, such as the shock of a loved one's death, or chronic, like grinding daily routines that frustrate basic needs. From degradation at Abu Ghraib to deprivation in Harlow's monkey experiments, a nasty enough environment will eventually provoke signs of depression.

Yet the puzzling reality is that human depression is increasing in an era when environmental conditions are relatively benign.[1] The average citizen in Western society now lives longer, is less likely to starve, and enjoys considerably greater wealth than his sixteenth-century counterpart. Presumably these objective conditions for survival and reproduction would cause depression rates to fall, not rise to nearly one in five citizens. This environment-depression disconnect seems less strange when we appreciate that there are additional human-specific routes into depression. *Homo sapiens* has the dubious distinction of being a species that can become depressed without a major environmental insult.[2]

There is no scientific consensus about why human depression rates are rising in the industrialized world, but several compelling possibilities exist. Their common thread is our species' unusual relationship with mood and the doors it opens for unique routes into depression. A chimpanzee is capable of feeling bad, but only a human being can feel bad *about* feeling bad. Former tennis great Cliff Richey, in his memoir *Acing Depression*, described how he became engulfed by low mood: "One of the horrible things about depression—in addition to the foul, odorous, sick, deathly mood you're in—is that you're now spending so much of your time, almost all of it, just trying to fix yourself. You're consumed by, 'How can I fix this horrible thing?'"

Humans have a host of unique thoughts and reactions to low mood, many of which are highly cognitive. Only a human can keep a mood diary or write a book about depression.

We often think of what's uniquely human as uniquely better. Surely pride may be a reasonable emotion for the species that harnessed fire and put a man on the moon. It's easy to see traits such as advanced language, the ability to be self-aware, and participation in a rich shared culture as unalloyed virtues. Yet when it comes to "fixing" mood, all of these special human assets can turn into liabilities, with the unintended consequence of making depression worse.

Sinking Through Thinking

A hallmark human response to low mood is to try to explain it—as we do with moods generally. We use language to construct theories about where painful feelings come from. The basic idea is, "If I understand why I feel bad, I will know how to fix it." This impulse makes sense. It fits with a main function of low mood, which is to help draw attention to threats and obstacles in unfavorable environments. In a low mood, behavior pauses and the environment is analyzed more carefully.

However, exactly what "analyze more carefully" means depends on which species is doing the analyzing. The schnauzer, Ollie, just separated from his sister, may sit at the window for hours looking for signs of her return. Visual search is the sum total of his environmental analysis. When a human pines for a loved one, say a mother missing her son away at summer camp, the analytical field is far wider. Our outsized language capability draws in thoughts linked to the situation: "That head counselor seemed awfully young."; "Did I remember to pack sunscreen?"; "I wonder why we haven't gotten a postcard?" These thoughts may then trigger further mental images—a flash of Tommy drowning, the funeral—as well as feelings—a pang of guilt for ever letting him go to Camp Meadowlark in the first place.[3]

Such reflections on mood have a purpose beyond self-flagellation. The mood system is practical and most interested in what to do next, in finding the action that will enhance fitness. What people brood about is not random but tracks key evolutionary themes (finding a mate, staying alive, achieving status, defending kith and kin, etc.). Mothers and fathers worry about their children at summer camp because mistakes in child rearing are evolutionarily costly. A mother who figures out that she's dwelling on a failure to pack sunscreen can send a remedial Coppertone care package, and, the next time Tommy is sent away, he's more likely to be fully provisioned. Even the most backward-looking counterfactual thinking (coulda, shoulda, woulda) has a forward-looking element: understanding why bad things happened helps us prevent their recurrence.

Reacting to low mood with thinking has evolutionary logic; it enhances survival and reproduction (fitness). Sadly, what's good for fitness is not necessarily good for happiness. Only *sometimes* does thinking about mood enhance happiness. We see this fairly reliably in certain brands of psychotherapy, in which the process of thinking about mood and discovering its meanings is specially structured and guided by an expert.[4] For a novice to think his or her way out of low mood and depression to get to a happier place—that's a dicier proposition.[5]

Humans are understandably confident when trying to think our way out of a low mood. We solve so many other problems by thinking, such as how to get a stalled car to start or how to make a healthy meal out of scraps in the fridge. Becky, a college professor in Maryland, organizes a team to analyze old production data from a distillery to figure out the determinants of good whiskey quality and use this information to ascertain why the distillery's product loss between brewing and bottling is nearly twice the industry standard. She is now in an episode of depression. Every morning Becky wakes up and says to herself, "What can I do today to solve this problem?" But even with a PhD degree, considerable insight, and bookshelves filled with self-help books, her depression hasn't budged for thirteen months. If you speak with her, even in her depressed state, it is immediately obvious that she is intelligent. On paper, she has every reason to believe that she can solve her depression.

Yet most humans, including Becky, are not nearly as good at this as they think they are. And our confidence in thought makes it more difficult to recognize when thinking is not working. The pitfalls of such an approach are under-appreciated. In fact, "thinking your way out" might actually provide new ways *in*, new ways for low mood to deepen into serious depression.

Our advanced language and ability to hold ideas in mind, called *working memory*, combine to create a formidable meaning-making machine. Yet this machine can be too pro-ductive for our own good. It can easily churn out new in-terpretations of a troubling situation well after the situation has passed. On Friday, a worker can still be mulling over her boss's hostile comment from Monday and wonder, "Maybe it was that e-mail I sent three weeks ago that set him off." Once the meaning-making machine is in overdrive, a bad mood can prompt a potentially unlimited number of implications. We can generate dozens of seemingly plausible environmental reasons for the question: "Why I am so blue?" (My job is boring. I need to lose weight. We can't stop global warm-ing.) Even if you are feeling only a tiny bit sad right now, take sixty seconds to try this yourself. I doubt you will draw a blank on possible reasons. You'll have leads. Yet many of the leads will be false, that is, irrelevant to the real source of the mood. When the real source of low mood is a thyroid deficiency or a low-grade infection, an analysis of the envi-ronment is moot. Or worse than moot, because with all the attention we pay to the false leads (all the reasons I hate my job), we may find fresh reasons to feel low. The generation of false leads may be good for fitness (the value of an exhaustive search), but it's not always so good for happiness.

Given our natural reliance on and our confidence in thought, the urge to repetitively think about the causes and consequences of low mood can harden into a habit. Researchers label this habit of thought *rumination*. Some people enter a ruminative mode even when facing minor troubles, or even when their environment is benevolent. A consistent body of data—much of it collected by the late psychologist Susan Nolen-Hoeksema—shows that this is a dangerous habit. People who report a greater tendency to ruminate on a short questionnaire have longer periods of depressed mood in everyday life, are more pessimistic about the future, and have a harder time recovering from the effects of stressors such as a natural disaster or a recent bereavement.[6]

The human meaning-making machine is so good at what it does that it can generate interminable interpretations. When persistent thinking gets stuck, it does not arrive at a stable theory of the problem, does not solve it, and cannot come to terms with it. Far from engaging in active problem solving, a person may simply perseverate on the fact of the problem (or problems) for months on end.

When the meaning-making machine gets caught in this way, its analysis turns inward, shifting its focus from a problematic environment to a problematic self.[7] Analyses of various kinds of thoughts have found that those that repeatedly focus on the failings of the self are the kind most closely linked to depression.[8] Insistent problem solving by itself is not necessarily harmful. In fact, therapeutic techniques that bolster active problem solving (say by breaking a problem down into structured subparts) can be helpful for depressed persons.[9] It's the deconstruction of the self that really causes trouble.

As *Homo sapiens sapiens*, we know, and we know that we know. An elaborate conceptual self—another thing that's usually a point of pride—becomes a vulnerability. We're committed to our autobiographical self, our story. It's as if we have films of our own lives playing in our heads, with us cast as the heroes. Depressed people, however, recast their movies with themselves as villains and play them in an endless loop. A depressed chimp, lacking a deep autobiographical self, is spared this screening and will never have the experience of lying awake at night thinking, "I am a terrible mother." Our capacity to dwell on our own failings makes us more vulnerable to depression than our fellow mammals.

Humans also have a special category of failings because of our heightened ability to self-monitor: our failures to change mood. This was true for Becky, who said of herself, "As a goal-oriented person, I keep looking for (and trying) things I can do to snap out of the depression. Medication, meditation, sleeping pills, trying to spend time doing 'things that bring me joy' (which just backfires, because I end up feeling hopeless while I'm doing them)." Every day that the depression goes on, failures to change mood turn into nagging thoughts: "Why can't I just get over this?"; "Why am I so weak?" These self-monitoring statements become further fodder for rumination, which becomes further fodder for depression, and we are reminded once again that our powers of language are a decidedly mixed blessing.

As you can see, our interpretations of sad mood are powered by a meaning-making machine that is not easy to downshift. This explains why the most useless pieces of well-meaning advice to give someone in midst of a deep depression are "Snap out of it" or "Stop thinking about it." This advice is nearly impossible to implement; about as futile as asking a burn victim to stop feeling pain.[10]

Given the difficulty of squelching mood-relevant thoughts, different schools of therapy have discovered clever tactics to subtly tweak the dials of the meaning-making machine. One good example, shown in Figure 6.1, is an exercise called *mindful observation of thoughts*, which is the type of exercise that one would see in a more elaborate therapy called *mindfulness-based cognitive therapy*, which has been shown to have some utility in preventing depressed moods from spiraling downward.

This exercise is designed to help a person notice and acknowledge thoughts without becoming embroiled in them. In part III of the exercise, the task is to imagine the mind as the sky, with the thoughts in the sky as clouds rolling by. No doubt this requires practice! If you can do this exercise successfully, congratulations; you have learned some of the skills involved in becoming a detached spectator of your own mind. Once these have been mastered, the nastiest cognitions now have the same status as other objects in the world—puffy clouds—and no longer form the sole basis of reality.

A common characteristic of the therapies that slow the progress of low mood into depression is that they turn down the volume on the verbal meaning analyzer. Acceptance, mindfulness, or cognitive behavioral—all of these therapies involve building up the ability to tolerate negative thoughts as a means to break the cycle of thought-based rumination and close off this special human route into depression.

FIGURE 6.1. Mindful Observation of Thoughts.

PART I

1. Sit comfortably, preferably with your back straight and feet flat on the floor.
2. Close your eyes and continue to sit, quietly.
3. Focus your attention on the flow of your breath as it moves in through your nostrils into your body, back up, and out through your nostrils.
4. Keep your attention on your breath.
5. Stay with your breath for five to ten minutes.

PART II

6. Focus your attention on your breath until you feel settled and centered.
7. Now focus on your mind.
8. Observe your thoughts arising and departing, one after the other.
9. Notice how your mind engages the thoughts and one thought gives rise to another, and another.
10. Do not try to stop the thoughts. Just observe the play of the thoughts in your mind.

PART III

11. Focus your attention on your breath until you feel settled and centered.
12. Imagine your mind as the blue sky, with clouds rolling by in it.
13. When you observe a thought arising, imagine the thought as a cloud passing by.
14. Observe your thought clouds drifting through the sky and out of your attention.
15. Do not respond to the thought clouds; just observe.

Adapted from "Can Adult Offenders with Intellectual Disabilities Use Mindfulness-Based Procedures to Control Their Deviant Sexual Arousal?" by N. N. Singh, G. E. Lancioni, A. S. W. Winton, A. N. Singh, A. D. Adkins, & J. Singh, 2011, *Psychology, Crime and Law*, *17*, pp. 165–179.

The Perils of Persistence

If you have suffered from depression or known someone who has, you know that when people are really depressed, they have a strong urge to stay in bed. Mimi Lopez described her bouts, which typically last from two to five hours: "When in bed I zone out and it is either just blankness, a nothingness that runs through my head or one phrase just keeps repeating over and over which is: 'What am I going to do now?' I never get a response or solution to the question."

Why do depressed people lie in bed? It's not because it's great to snuggle under the blankets; it's because they can't bring themselves to get *out* of bed. Almost any other activity or task becomes a painful ordeal, even activities as simple as taking a shower or getting dressed.[11] This seems strange. A perfectly able-bodied person can't bring herself to rise out of bed. How does this happen?

The intuitive answer is that this reflects a lack of motivation. Depressed people are directionless because they are undercommitted to goals. Without goals to drive future behavior, current behavior becomes frozen for long periods. Bed is the most natural location for a behavioral pause, as the place in the house most associated with inactivity.

The intuitive answer is okay as far as it goes. The problem is that it just doesn't go very far. It raises the further question of how a person loses the desire to pursue goals in the first place.

The alternative answer involves a surprising theory about another uniquely human route into depression. Recall that low moods tell us when our progress toward goals is poor.[12] Often low moods arise when we hit an obstacle. Our usual first reaction is to redouble our efforts to achieve the goal. If the goal still proves unreachable, the low mood will escalate further. At some point, however, something has to give. Typically, as low mood escalates, the person will give up on, or scale back, the goal and/or move on to another activity that has a better payoff.

Eric Klinger and Randolph Nesse have argued that low mood has utility in helping animals disengage from futile efforts.[13] In a world where time, resources, and effort are precious and finite, having an evolved mechanism to hasten disengagement from a failing goal is important to survival. If it remained at a barren fishing spot, our bear at the river bend would starve.

Humans have a decidedly more complex relationship with mood. We can choose to act on mood or ignore it. People can try to override mood and continue pursuing a failing goal, like firewalkers who forge on despite the dangers of smoldering coals.[14]

What results is a standoff between the imperatives of a person and an ancient mood system. To resolve the standoff, the mood system must do something drastic. It turns down the volume on goal pursuit, not only on the one goal, but on goals across the board. Eventually, when the mood system wins, it can result in flat-on-your-back depression, with fatigue, torpor, and lack of motivation—the whole nine yards.[15] Hence the remarkable phenomenon of a person lying in bed and staying there—with all other options for actions to pursue shut down and rejected as less tenable.

So this alternative theory turns the standard explanation on its head. Depressed people don't end up lying in bed because they are undercommitted to goals. They end up lying in bed because they are overcommitted to goals that are failing.

The idea that depression results from an inability to disengage efforts from a failing goal is relatively new. Could it be a plausible pathway into depression?

Population level survey data suggest an affirmative answer. More people in the West—especially the young—are setting the kinds of goals that are likely to become failing goals in the future. From 1976 to 2006, the percentage of high school students who said that having a lot of money was "extremely" important rose from 16 to over 25 percent.[16] By 2000, half of high school seniors reported they were aiming for graduate school (law, dental, business, etc.); this rate of aspiration had doubled from the 1970s, yet over the same period, the odds of a high school graduate finishing graduate school were flat, under 10 percent.[17] In 2005, 31 percent of American teenagers said that they were going to be famous someday.[18] Even if we count appearances on reality TV as fame, only a tiny fraction will achieve their goal. Finally, over just ten years, from 1997 to 2007, the number of Americans undergoing cosmetic surgery jumped more than five times, a behavioral sign that the goal of extreme beauty is rising in salience.[19] Today's youths are also more likely than those of the past to agree with statements such as, "I will never be satisfied until I get all that I deserve,"[20] which suggests that these goals are more than idle fantasies. Incidentally, you may consider goals like wealth, fame, and beauty morally suspect and feel glad (on some level) to hear that these kinds of extrinsic goals are associated with lower levels of well-being and happiness.[21] But here the content of the goal is not so important. Commitment to *any* immodest goal—it could be world peace or unconditional mother love—can fuel depression.

Additional evidence for the overcommitment theory is that perfectionists are more likely to become depressed than nonperfectionists. Perfectionism involves a tendency to maintain high self-expectations about goal completion.[22] This was how Maria described herself: "Most people who know me well say that I am too driven at times and don't give up or 'give it a rest.'" She continued with a recent example:

> When I bought an old house that needed to be restored I worked full time, had two young children, and all my free time was spent restoring the house. I had worked overtime on a Saturday, came home around 4 P.M., made dinner and went to wallpaper the bedroom. I was up until 4 A.M. finishing the project I had started. I never like to leave something undone, unfinished. I will work on it until I am satisfied with what I set out to accomplish. Why do I do this? Something eats at me until I finish. My focus is so intense.

One can easily see how this admirable quality — persistence — could become a liability — an inability to disengage — when Maria's efforts hit a wall.

The idea that depression is about holding onto failing goals also fits clinically with the kinds of situations that often precipitate serious depression: the battered wife who cannot bring herself to leave her troubled marriage, the seriously injured athlete who cannot bring himself to retire, the laid-off employee who cannot bring herself to abandon her chosen career despite a lack of positions in the field.

Evolution sculpts the broad terms of what we are motivated to commit to. Goals that will bring us status, security, power, allies, attractive mates, and the like tend to be compelling. The human meaning-making machine is charged with filling in the details, with interpreting what the most important goals in any given environment are. The room for interpretation and the range of human enterprises help explain why your priorities are probably different from those of a Kalahari Desert tribesman. The constraints of evolution still apply, of course, even as humans pursue goals more diverse than those of tigers or tree shrews.

What may be most important for exposing humans to the risk of depression is that they are able to pursue highly abstract goals and to set goals in domains where progress is difficult to measure.[23] The story of James Stimpson is illustrative:

> *I'm 57; have been depressed most [of] my life. My problems begin in childhood. My father was a monster, clearly a psychopath. Only my mother could not recognize him for what he really was. She was blinded by denial, with a pollyannaish view that "things will get better tomorrow."*
>
> *This expectation for a better tomorrow clashed with an awful reality: Things just kept getting worse. So it goes with psychopaths, who become better practiced at manipulative games over time. My mother kept pushing on me that, someday, I would win my father over. I tried to earn his pride, culminating in an Army hitch (where I performed well). But to no avail. It eventually became undeniably clear, at least to me, that whatever I did, or did not, do would have no effect.*
>
> *My mother never forgave me for having "given up" on my father. This left me with a one-two punch: My father*

consistently pronounced me worthless, I let my mother down because I failed to turn my father around. I disappointed both parents. This left me with a sense of failure from which I've never recovered. The most important challenge of my life . . . [a]nd I fell short.

In late middle age James eventually came to recognize that his commitment to a failing goal was a major theme in his life and was the main reason he had experienced chronic problems with depression. His goal was the fundamental one of winning the love of his parents, but especially that of his father. His story dramatizes how goal pursuit controls mood and how difficult it can be to detach from failing goals.

All Cheered Out:
Culture and the Pursuit of Happiness

We are the only species to look to culture to guide us on what feelings are desirable and how undesirable feelings should be managed. And as humans try to "fix" low mood, they are never alone. No creature ever living has had available so much advice—spiritual, medical, psychological, folk-inspired—about what to do when it's feeling down. In the past fifteen years we have seen an ever-growing stream of psychological and popular science books examining happiness and how people can increase it (see Figure 6.2).[24] Ideally, these resources should serve as bulwarks against depression. Perversely, the opposite may be the case. Our predominant cultural imperatives about mood, though surely well-intentioned, are worsening the depression epidemic.

FIGURE 6.2. American Culture of Self-Help Feeds High Expectations for Happiness.

Photo credit: Ena Begovic

In the West there is a powerful drive to experience happiness. This tradition is particularly strong in the United States. Indeed, it's difficult to think of anything more American than the pursuit of *happiness*. Along with life and liberty, it's written into the Declaration of Independence as a fundamental right. Wanting happiness is as American as apple pie. But how happy should we expect to be? Happier than other people around the globe?

It would appear so. Analysis of thousands of survey responses found that when people in different countries were asked to rate how desirable and appropriate it is to experience varying psychological states, positive states like joy and affection were rated more desirable and appropriate in

Australia and the United States than in Taiwan and China.[25] Cross-cultural research by Jeanne Tsai of Stanford University has also found that European Americans place the highest value on specific forms of happiness, idealizing states like enthusiasm or excitement, termed *high arousal positive states*. By contrast, Chinese and other Asian test subjects place the highest value on other forms of happiness, idealizing states such as calm and serenity, termed *low arousal positive states*.[26]

Consistent with the notion that culture inculcates ideals about feeling states, cultural differences show up early in life. When young children judge smiling photographs, American children prefer the expression that shows an excited smile to the expression that shows a calm smile. Taiwanese children do not show this same preference. American preferences for high arousal positive states probably have many roots, but they stem in part from a media environment that values peppy happiness. An image analysis of smile photos in American women's magazines found that they contained more excited smiles and fewer calm smiles than smile photos in Chinese women's magazines.[27]

So what's the problem? Everyone I know wants to be alive, free, and happy. What's wrong with pursuing happiness to the fullest extent possible? The more you value your happiness, the happier you'll be, right?

Wrong, says compelling recent research.

Two studies led by psychologist Iris Mauss found evidence for an alternative hypothesis: people who value happiness more are *less likely* to achieve their goal of feeling happy. In the first study the researchers administered a questionnaire designed to measure the extent to which people valued the experience of happiness as a fundamental goal. Mauss

FIGURE 6.3. Women Who Value Happiness More Report Lower Well-Being During Periods of Low Stress.

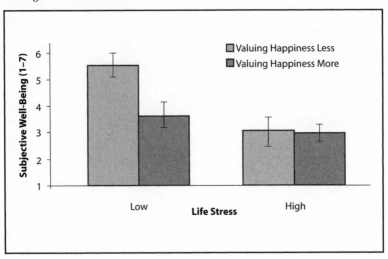

Adapted from data reported in "Can Seeking Happiness Make People Unhappy? Paradoxical Effects of Valuing Happiness," by I. B. Mauss et al., 2011, *Emotion*, *11*, pp. 807–815.

and colleagues found that some people put an especially high value on happiness, endorsing items like, "If I don't feel happy, maybe there is something wrong with me"; and "To have a meaningful life, I need to feel happy most of the time."[28] Surprisingly, women who said that they valued happiness more were actually less happy than women who valued it less. Specifically, women who valued happiness highly reported that they were less satisfied with the overall course of their lives and were more bothered by symptoms of depression. Strangely enough, valuing happiness seemed most problematic for women whose lives were low in stress—the people for whom happiness should have been within easiest reach (see Figure 6.3).

The second study was a clever experiment in which the researchers tried to briefly increase how much the participants valued happiness. They did this by having one group of participants read a bogus newspaper article that extolled the importance of achieving happiness (the other group read an article that did not discuss happiness). Later in the experiment participants watched different short films. Those women who had read the happiness-extolling article reported feeling less happiness in response to a happy film. The authors again concluded that, paradoxically, valuing happiness more may lead people to be less happy, especially when happiness is within reach.

These experiments help us understand why predominant cultural imperatives about mood might be worsening the depression epidemic. Our current cultural ethos is that achieving happiness is like achieving other goals. If we simply work hard at it, we can master happiness, just as we can figure out how to use new computer software, play the piano, or learn Spanish. However, if the goal of becoming happier is different from these other goals, efforts devoted to augmenting happiness may backfire, disappointing—and potentially depressing—us because we can't achieve our expected goal. Mauss and colleagues concluded that setting a goal to become happier is like putting yourself on a treadmill that goes faster the harder you run.[29]

Rising happiness standards widen the gap between what we want to feel and what we actually feel. We know from Jeanne Tsai's work that people in the West generally idealize excitement and other high arousal positive states. Although this is a general tendency, she has also shown that people vary in how strong their positive ideal is. Importantly, for people who have that strong positive ideal, there is potentially a large gap between what they would like to feel and what they actually feel. The size of this gap predicts depression. People who have a larger gap between their ideal and actual positive affect have more depressive symptoms.

This is not surprising: to someone with high happiness goals, low moods are as demoralizing as a foreclosure notice is to an aspiring billionaire. If you believe that a high positive mood should be easy to achieve, a prolonged low mood is an insult, which probably prompts the isolating and stigmatizing question: "What's wrong with me?" Negative feelings about negative feelings make them a greater threat. People who set unrealistic goals for mood states may be less able to accept or tolerate negative emotional experiences like anxiety or sadness.[30] Oddly enough, being able to accept negative feelings—rather than always striving to make them disappear—seems to be associated with feeling better, not worse, over the long run. There is evidence that when people accept negative feelings, those experiences draw less attention and less negative evaluation than they would otherwise. Some research shows that people who report an ability to accept negative feelings when they arise are less likely to experience depressive symptoms in the future.[31]

Ultimately, the strong cultural imperative toward being happy bumps us up against a wall: our mood system is not configured to deliver an end state of durable euphoria. Happy euphoria is a reward the mood system metes out along the way, on the road to pursuing other evolutionarily important goals. For example, euphoria is a reward for having sex or for when your first-choice date to the prom says yes. By design, these rewards are meted out sparingly rather than liberally.

Yes, pleasure after eating the carrot rewards the bunny for finding the carrot, yet a well-designed bunny does not stay satiated. It's the *end* of the pleasure and the promise of *more* that gets the bunny hopping off to find more carrots and ultimately to survive long enough to make more bunnies. So clearly does intense happiness fade after a goal is achieved that psychologists and economists have given the experience its own label: *hedonic adaptation.* It is powerful, and studies show it to be virtually omnipresent: whether after purchasing a zippy new sports car, getting a big promotion, or moving to a cool new apartment, with time (often surprisingly little) the euphoria fades.[32]

Hedonic adaptation and our unattainable cultural imperatives make for a cruel combination. People will usually fall outside the zone of intense pleasure, and they will consider that failure. In this scenario, shortcuts are tempting. Forget having to realize an evolutionarily important goal and just give me the pleasure now, please. The high from smoking crack is almost immediate, but it does not last. In the long run, the shortcuts backfire. The mood system has the last word.

Now What?

Because human depression can happen without a major threat to fitness, it's more complicated to unravel than rat depression or dog depression. Of course human depressions can be relatively pure fitness events, determined by when the environment lands a body blow. However, the reaction to such blows is not always predictable. On the theme of finding a mate, a woman might become depressed soon after her husband announces he's not attracted to her anymore and wants a divorce. For another woman, because of how she manages mood, a minor blow—rejection after a first date—is enough to get depression going. This is because the bad date sparks low mood, which sparks obsessive thinking about reasons for romantic failure, which over time deepens low mood, which, with the frustration of strong expectations for happiness, deepens low mood still more. Soon her depression is just as deep, just as *real*, as if the bad first date had been a divorce filing. For her, and people like her, depression does not need to wait until a grave challenge to fitness is posed. Minor precipitants are enough.

To this point I have laid out why the mood science approach has unique power to explain the origins of the depression epidemic, as well as why depression can be so tenacious when it takes root. Depression draws sustenance from many admirable human qualities, such as our ability to think and use language, our tendency to hold onto ambitious goals, and even our drive to be happy. The picture of depression that emerges is richer, more interesting, and in some ways more troubling than defect-model approaches would allow.

We have seen the remarkable confluence of unfortunate circumstances, some originating millions of years ago, that have come together in a "perfect storm" for mood. The depression epidemic has emerged from the ways that our evolved capacity for mood plays out in the convergence of particular temperaments, routines, and stressors with contemporary strategies for managing mood. It is chilling to think that the human depression epidemic results less from defects than from what humans do too well, which can help to explain why depression has become such an entrenched problem, one that won't be eradicated soon.

This is a somber argument, and you may wonder if the mood science approach is unduly pessimistic. I don't see it that way. Rather, I see it as delivering a dose of much needed realism in our dialogue and analysis of depression. We will not be able to reduce the reach of depression until we have a sober understanding of what we are up against. At the same time, it must be granted that it is difficult to translate our approach into a three-step program.

Nevertheless, rejecting easy answers is not the same as giving up; far from it. Even though it is not an advice primer, this chapter does suggest clues about how low moods can be better managed before they turn into severe depressions. For example, it offers a fresh appreciation of the costs of thinking and the benefits of becoming a critical consumer of your low mood, even sometimes accepting a low mood with equanimity; why goals should aim high, but not too high; and why we should recognize that persistence in striving to achieve a failing goal can be self-defeating. It also suggests that we should avoid fixating on a specific happiness level and recognize that happiness itself is not a goal but a fleeting by-product of progress toward other goals. As we move forward and consider the often unremitting toll of severe depression, the mood science approach—by its integrative nature—will continue to suggest clues for the management of mood and which levers we might pull to move mood into a better place.

CHAPTER 7

The Black Hole:
The Psychobiology of
Deep Depression

SYLVIE DESCRIBED THE WORST POINT OF HER DEPRESSION AS "WALKING around in a concrete dress." She elaborated, "My joy was gone. I stopped listening to music. I stopped putting down the top on my convertible. Driving became hard. I had to concentrate and go slow. My arms on the wheel were heavy. I could not even cry. I was flat. I could not make or tolerate conversation. My answers were yes, no, I don't care, Sure, whatever you want, I can't decide, you decide, it doesn't matter."

Deep depression destroys the ability to *do*. It renders simple tasks impossibly difficult. Author Jeffrey Smith described his depression as a kind of suspended animation. Every night he would lie in bed for hours, exhausted, in a holding pattern, "unable to return to sleep, but neither could I will myself out of bed." When day finally came, the simplest acts of the morning routine were an insurmountable challenge. He wrote, "When it was time to ready myself for work, the stairway to the kitchen and bathroom seemed too steep to climb, and then my arms felt too heavy to lift and wash myself in the shower. So I just stood there under the water."[1]

In this strange landscape, losing the ability to move, converse, sleep, or read is common. Depressed people are often apathetic. They scale back their ambition and interests; they lose interest in group activities; they feel older; they perceive themselves as ineffective.

There are many things that depressed people cannot do or cannot do well. For defect models, this is proof enough that depression is pointless.

The logic is simple: remove the underlying defects—be they low serotonin or pessimistic thinking—and wipe depression from the human repertoire. On its face, this is a compelling argument. We live in a culture that values doing. As members of this culture we are predisposed to view conditions in which people cannot *do* as diseases. It is difficult to understand how "walking around in a concrete dress" could be useful or to recognize situations in which the capacity for depression, even deep depression, might be necessary to survive.

Yet our capacity for deep depression is an evolved response, and one with a purpose: to organize disengagement. Recognizing this purpose helps us understand why depression exists in the first place, despite its obvious costs, and why severe depression can be so terribly persistent.

As WORLD WAR II neared its end, thirty-six idealistic conscientious objectors, members of the Civilian Public Service, volunteered to be systematically starved. The project, headed by Dr. Ancel Keys, was designed to develop an understanding of the physiology and psychology of starvation and to create strategies to manage the mass starvation that might follow the war's end in Europe. In the major phase of the experiment the men received restricted rations (about sixteen hundred calories a day) for six months. Their dietary compliance and health progress were repeatedly monitored in a state-of-the-art lab underneath the University of Minnesota football stadium. Meticulously recorded in a two-volume tome, *The Biology of Human Starvation*, the experimental protocol had the men lose on average 25 percent of their body weight.

Starvation did more than shrink the men's bodies; it took a toll on their minds. The men had initially been selected for their good psychological and physical health. Morale was high at the start: this was a history-making experiment and part of a heroic war effort. However, starvation changed vital, healthy males into men haunted by signs of depression. In addition to slowed metabolism with reduced heart rate, they experienced low energy, at times to the point of being unable to move. When the effects of starvation on mental abilities were tested, the men struggled to concentrate on the tasks at hand. Their thoughts and attention kept going back to food, with an obsession so strong that one man reported dreams of cannibalism. The men completely lost interest in sex. And for most, starvation brought with it prolonged low mood, recorded in diaries and on scales of psychiatric functioning administered during the experiment. In other words, the volunteers showed many of the signs of significant depression.

FIGURE 7.1. Starvation and Deprivations
Such as Those Experienced in the Civil War
Prison at Andersonville, Georgia, Are Strong
Cues for Depression.

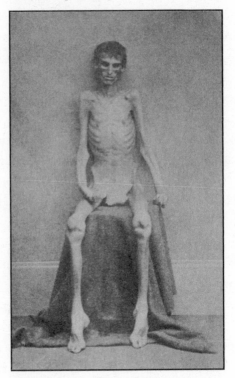

Photo credit: US Library of Congress

One lesson is that for some emergencies, the best survival
strategy is to hold in place. The Minnesota Starvation Experiment simulated famine and extreme food scarcity, situations
that repeatedly killed our mammalian forebears and even
now threaten many millions of people.[2] Animals that reacted
to famine with high mood and bold new ventures were less
likely to make it through than those that responded with low
mood and behavioral withdrawal (see Figure 7.1). As descen-

dants of those who made it through, evolution has inscribed in our minds an advance directive for behavioral shutdown.[3]

Does the example of famine mean, then, that deep depression is physiologically induced? The answer is not quite so simple. We also carry this advance directive to prepare us for psychological situations that repeatedly threatened our forebears. The strongest depression-inducing situations present a double whammy: serious losses and no route (or an overly hazardous route) forward. These situations are akin to psychological starvation.[4]

It also bears repeating that just because the capacity for depression is an adaptation does not mean that episodes of depression are always good or useful. Chapter 6 showed how human cognitive and cultural capacities can subvert what nature intended and turn on a depression even in the absence of a major stressor. Even when depression is activated at the right place at the right time, it carries significant costs. Given the costs, it stands to reason that this response should be used sparingly.

Emotion Context Insensitivity

Poor functioning in deep depression tells us about depression's object. Deep depression is an organized response that arises to make sure we don't act. The mood system seizes the entire body and mind, every drive in the body—eating, sleeping, fornicating, or emoting—and pulls them toward disengagement. In cosmology, a *black hole* is an invisible body that is created from a collapsed star so dense that neither light nor matter can escape. In severe depression, the mood system is much like a black hole, exerting such strong gravity on motivation that people stop doing; otherwise able-bodied people cannot lift themselves out of bed. The idea of depression as organizing disengagement—a black hole for motivation—may seem strange or paradoxical. It was for me, at least at first. When I entered science this was not a common way of seeing depression. Only after a series of events—both professional and personal—did I arrive at the conclusion that it was the only theory that made sense.

———

ON A WARM LATE SUMMER DAY IN 1997, I was biking across the sprawling Stanford campus from Jordan Hall to my apartment in Escondido Village. As I pedaled through Palo Alto I flashed back to the turn of events that had brought me to study emotional reactivity in depression. Not too long before I had been irretrievably and irremediably lost in the black hole of depression, even an inpatient. Depression had had its way with me. I had gone from being a somewhat cocksure Ivy League grad who had sailed into a history PhD program and was banking on leading a life of the mind to someone who could barely read a grocery list. I went from fairly sure that I was going to be listened to, to completely convinced that I had nothing to say; that I was completely broken; and that the last chapter of my life had been written in stone and would end with a loss of mind, loss of station, and my total obliteration as a human being.

After a year or so at this rock bottom, I had the wherewithal to take some courses at a local community college—in psychology—with the vague sense that therein might lie a new path. Though I learned little in my compromised state, I learned enough to make a decision: I was going to understand how mood could overwhelm. I was going to understand depression or die trying. At the least I was going to find out if anyone would take a chance on a failed historian.

As I pedaled through the palms, I remembered hearing Laura Carstensen's smiling voice tell me that this was her favorite part of her job, then hearing her offer of admission into Stanford's PhD program. I had hung up the phone, started crying, and hugged my wife. I had been given a chance to study emotion with Dr. James Gross and depression with Dr. Ian Gotlib, both generous mentors and world-class scholars. Working with James and Ian, I might win back my life and create lasting science in the process. I could not let them down. I could not let myself down. This was a chance at redemption.

Walking on this new path, I slowly found my feet.

In my new field the question I pursued had both basic research and applied import: *How does a sustained mood disturbance influence ongoing emotional reactivity?* As I dug into previous work, it seemed that the answer varied depending on what type of emotional stimulus was involved. For positive emotional stimuli like a beautiful sunset, the consensus was that depressed people should be less engaged, less reactive. For negative stimuli, most opinion was in the other direction, that depressed people should be *more* engaged and reactive. In fact, despite few tests of this idea, a wall of opinion held that depression enhanced reactivity to negative stimuli.

One brick in this wall was from emotion theory.[5] Here the default view was that moods amplified an emotion when the mood and the emotion were similar in nature. For example, an anxious mood should make you more likely to startle in fear at an unexpected sound. By the same logic, a garden-variety sad mood would enhance your reaction to hearing tragic news. And by extension, a clinically significant depressed mood should really boost responses to all things sad.

Cognitive theories of depression were a second brick in this wall. These theories also predicted that depressed people would be highly responsive to negative stimuli. Psychiatrist Aaron Beck proposed a core idea that people are vulnerable to depression by virtue of problematic cognitive structures called *negative schemas*, which guide how we process and react to our environment. These negative schemas are a hidden force that shapes our explicit beliefs ("I'm unlovable") as well as what we pay attention to and what we remember. (Do sad memories come more easily to mind than happy ones?) These structures are inactive, causing no problems for a vulnerable person until a negative mood develops. In Beck's view, negative mood strengthens the negative schemas. As they strengthen, they become more attuned to negative stimuli in the environment. The schemas respond to any negative stimulus that matches them, leading to vigorous emotional reactions that sustain depression.[6]

Finally, the third brick in the wall was observations of depressed people drawn from clinical settings. A prime example was the DSM manual, which in addition to listing the formal diagnostic criteria for depression contains "liner notes" on characteristic qualities of mental conditions based on accumulated clinical wisdom. One of these notes states that depressed people are highly prone to crying spells. These are thought to be so typical of depressed people that a clinician can consider crying in a patient a symptom of depression. To the extent that these observations are correct, they also suggest that the depressed are an emotionally volatile group who are easy to perturb.

When I looked at this wall a bit more carefully, however, it didn't appear so solid. It bothered me that strong opinions had formed about emotion without much controlled research. Ian and James were both master experimentalists who taught me the experimentalist's credo: theories without data are like daydreams.

It struck me that our prevailing assumptions about depression might not survive the light of experimental data. Depressed people generally report feeling high levels of sadness, anxiety, or anger. Do these feelings reflect innately high emotional reactivity? Maybe. Alternatively, couldn't such feelings just as easily reflect lives jam-packed with negative stimuli? Wake up to rejection by a hostile spouse, drive to a dead-end job, and go home to a stack of unpaid bills. Sleep. Repeat. By temporarily taking people out of their sometimes chaotic lives, the beauty of an experiment lies in its objective test of reactivity—in an experiment the emotional stimuli are exactly the same for everyone, and they are presented in a controlled laboratory setting.

Likewise, I didn't entirely trust the DSM's observations about increased crying in depression. When a person discloses painful experiences to a sympathetic therapist, these might not be ordinary moments, but occasions that invite crying. (Consultation rooms are equipped with tissue boxes for a reason.) Without a wider range of observations, I wasn't convinced that depression lowers the threshold for crying. What would happen, I wondered, if we brought depressed people into a lab setting and tried to provoke them to cry? Would they be any different from psychologically healthy people? We needed to perform experiments to find out.

I admit that an experiment examining how easy or difficult it is to make depressed people sadder may sound sadistic. (I could almost hear the whispering: *Rottenberg, yes he's that fellow who makes the depressed people cry*.) But for the sake of science, it was critical to convince myself, my advisors, and the Stanford Human Subjects Review Board that the experiment was ethical. What we were doing was like a stress test, in which a cardiac patient runs on a treadmill while the heart's activity is carefully monitored. When stressed, the hidden defect in the heart's rhythm is revealed. So, too, by perturbing emotion in a controlled setting we could begin to accurately diagnose how depression alters reactivity. We had safeguards, including a psychiatrist on standby, in the event a participant became acutely distressed and inconsolable. The potential scientific benefits were clear: by studying mood and emotion in depression, we were focusing on the most important parts of the most important mental health condition. Given how precious this knowledge was, it would be unethical *not* to proceed.

To prepare for the first study, my colleagues and I edited short films designed to target specific emotions. Because no one elicits emotion better than Hollywood, we stitched together material from commercially available films designed to elicit specific emotional states. Of particular interest were movies edited to elicit either sadness or a neutral state (i.e., few reports or displays of emotion). The sad film dramatized a death scene and revolved around themes of loss and grief; the neutral film depicted relatively innocuous landscape scenery. We pretested the films with healthy populations to confirm that they were effective.

It was time to bring people in and let the data decide. The first stage was enrolling the sample, which required conducting detailed psychiatric interviews to ensure that participants either had depression or had no history of any mental condition. Then, on a separate occasion, we recorded depressed and nondepressed participants' self-reported emotional experiences and their observed expressive behavioral reactions (via an inconspicuous camera); finally, we monitored physiological reactions to the films, like heart rate, sweat gland activity, and blood pressure, via small sensors attached to their hands and bodies.

What did we find? First, contrary to clinical lore, we discovered that depressed people were not any more likely to cry at a standardized sad situation than nondepressed people. About one in five depressed people visibly cried, the same rate seen in the healthy subjects. And when depressed people did cry, their crying episodes were actually less vigorous. As shown in Figure 7.2, when reactivity was computed from a neutral reference point, depressed criers actually exhibited *smaller* changes in their emotional experience and physiology than healthy criers did.[7] In my first scientific experiment, depression appeared to lessen, not magnify, the crying response.

When we looked broadly at the data from this experiment — at emotion experience, behavior, and physiology — depressed people showed no evidence of exaggerated reactivity to the sad film. On the contrary, depressed participants reported greater sadness than healthy participants in response to the neutral film. When the neutral film was used as a reference (a typical practice in studies of emotion), the sad film led

FIGURE 7.2. Depression Blunts the Crying Response.

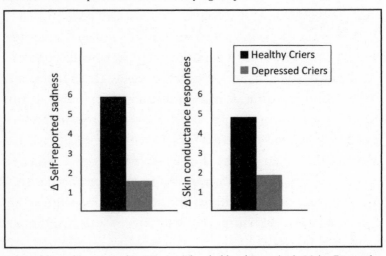

Adapted from data reported in "Crying Threshold and Intensity in Major Depressive Disorder," by J. Rottenberg et al., 2002, *Journal of Abnormal Psychology, 111,* pp. 302–312.

depressed subjects to report smaller increases in sad feelings than the healthy controls did. More, this lack of reactivity could not be explained by a ceiling effect—that is, it was not a consequence of depressed persons' sadness already being at the upper limit of measure while watching the neutral film. In fact, the difference remained significant even after removing from the analysis those depressed participants who had reported very high levels of sadness in response to the neutral film. Against the received wisdom that a sad mood would enhance reactivity to a sad stimulus, we found depressed persons' reactions to an acutely sad film were not much different than their reactions to viewing innocuous landscape scenery.

Were these first results a fluke? In my dissertation study, I intended to try to reproduce the results and clarify the interpretation. One potential problem with the original experiment was that our carefully edited and standardized Hollywood film, though undoubtedly sad, might be irrelevant to the particular concerns of a person suffering from depression. Watching the pain of another could even be a momentary distraction, a break from one's own nightmare. To remedy this issue we conducted another experiment containing sad stimuli that we were confident were highly personally relevant. We created these stimuli by videorecording the participants speaking about sad events they had experienced. Later we played back the video to participants (as well as recordings of participants relating happy events and neutral information). Nevertheless, even with sad stimuli constructed to be high in personal relevance, the depressed people still had reduced reactivity.[8]

By the time I moved to Tampa to take my first job as an assistant professor at the University of South Florida, other laboratories had been at work performing experiments with a wider range of emotional stimuli. With my graduate students, Lauren Bylsma and Beth Morris, I set out to compile and combine results from all the experiments, using a statistical technique called meta-analysis. The pooled data from nineteen laboratory studies had a combined sample size of over nine hundred participants. The meta-analysis also allowed us to pool data from a wide variety of emotional measurements taken during the experiments — from feelings participants reported, to facial expressions like furrowed brows, to bodily changes like heart rate or blood pressure reactions. Together these analyses would give us much more definitive and reliable results than any single study could ever do.

Lauren and Beth's analysis of reactivity to positive emotional stimuli came in as expected. When we aggregated across studies, depressed people were less reactive than nondepressed people. The stunning result was that the same pattern—less reactivity—also applied for the negative stimuli. The Stanford results held up across nineteen experiments. If anything, the pattern was now clearer. We dubbed this pattern *emotion context insensitivity* (ECI for short) to describe a loss of appropriate reactivity to emotional stimuli.[9] Depressed people were emotionally "stuck," reporting and displaying the same emotions in different situations. It was as if the stimulus no longer mattered: a beautiful sunset, a newborn baby, a flower arrangement, rotting maggots— everything was, equally, subpar.

———————

I DID NOT DISCOVER emotion context sensitivity. The phenomenon has always existed in depressed people. The problem was that a wall of conventional scientific wisdom had temporarily obscured our ability to see it. Yet the more I thought about it, the more ECI made sense; the more it seemed like an important truth that squared with other things we know about depression.[10]

For one thing, it fits with what depressed people say about themselves. Depressed people experience their present world as undifferentiated: life is flat, dull, empty, and unprofitable. Their future world is more of the same—an endless plain of misery. Our most vivid descriptions of severe depression portray it as a monotonous trap. Author William Styron, a famous depression sufferer, once remarked about its horrible sameness, "The depressed person is condemned to life."[11]

The pattern also fits with earlier observations of depressed people, often taken from inpatient wards. Patients are remarkably unresponsive in their facial expressions, and when left to their own devices, stare off into space with a fixed gaze.[12] At its strongest and purest, this becomes catatonic depression, during which a person may not move or talk for long stretches of time. Although this shutdown of dynamic behavior may be most dramatic in severely depressed inpatients, it also appears to exist in other settings and in less severely depressed persons. In 2010, Peter Kuppens and his colleagues took careful second-by-second measurements of behaviors indicating happiness, anger, and sadness during fifty-four minutes of interaction. For all of these behaviors, depressed people showed less moment-to-moment change in emotional behavior than nondepressed people.[13]

We might consider inflexibility a sort of master metaphor for serious depression. It is present not only in behavior but in cognition as well. I previously mentioned rumination as a kind of stereotypical thinking, in which people engage in circular, looping thoughts about the reasons for and meaning of their own sad, dysphoric affect. Not only is rumination an inflexible cognitive style, it also represents a passive, inactive mode wherein thinking displaces more active engagement with the environment. So, too, other cognitive researchers have identified other patterns of cognitive inflexibility, such as ways that depressed people lock themselves into specific kinds of explanations for life events (called attributions). When people inflexibly blame themselves when something bad happens, or always see bad events as having broad and long-lasting effects, this too feeds depression.[14]

In fact, the more places we look, the more we see evidence of depressive inflexibility; this includes the brain and other aspects of bodily functioning. Neuroimaging studies of deep depression show that depressed people have higher activity in brain networks that support introspective functions and lower activation in brain areas during tasks involving external attention to objects or events outside the individual.[15] Decades of research on the stress hormone cortisol indicate that many depressed people chronically overproduce this hormone. More revealing, the chemical switches that normally turn cortisol off appear to be stuck. Depression flattens out the normal daily rhythm of the hormone. In situations in which cortisol normally spikes—such as having to give an impromptu speech to a critical audience—depressed people don't exhibit a spike.[16] In depression, the body is equally at war whether it's lying in bed alone, being cradled by a lover, sitting in heavy traffic, or being subjected to verbal abuse.[17]

Is deep depression an organized response? Is it a legacy of evolution that serves to shut down motivation? Data over the past fifteen years (collected by me and by others) have led me to the point where I am convinced of an affirmative response to these questions. But if you are not yet there, another line of inquiry points to similar conclusions via a different route.

Why Does Deep Depression Last?

One of the hallmarks of deep depression is that it tends to extend over long periods. On average, an episode of major depression lasts about six months, with some episodes stretching out over years.[18] One question is whether the idea of organized disengagement helps us understand why depressions often endure. If a main point of depression is to organize disengagement, then one could surmise that the harder that motivation was shut down, the longer depression would go on.

I wanted to test this the-harder-the-shutdown, the-worse-the depression prediction. It seemed possible because some depressed patients showed the ECI pattern more than others. Would those depressed patients with the most pronounced ECI—the most depressive disengagement—also have the worst prognosis for their depression?

It appears that the answer is yes. In my original study we found that those depressed persons who reported the most similar reactions to sad and neutral films (the constricted pattern expected by ECI) had the worst functioning; their depression was highest in number and intensity of symptoms, their episodes had been going on the longest, and their psychosocial functioning was judged by an interviewer to be the poorest.[19] In a separate test we found that depressed individuals who disclosed the least sad emotion when discussing memories of sad life events also showed the least improvement in their symptoms one year later. A third test examined the predictive power of physiological reactivity. Here, those depressed people who watched a sad film without showing *vagal withdrawal* (a reaction of the autonomic nervous system we assessed with an electrocardiogram) were the least likely to recover from depression six months later. Other research groups are extending these results. In one particularly intriguing experiment, Peter Kuppens and his colleagues found that those adolescents who had less second-to-second change in emotional behaviors during an interaction with a parent would be more likely to have a first episode of depression in the next two and a half years.[20]

All of these data are an embarrassment for conventional wisdom. If the core problem in depression were exaggerated reactivity to negative stimuli, we would expect that those depressed individuals with the most exaggerated reactivity to negative stimuli would be the worst off and face the worst prognosis. Repeatedly, results have shown the opposite, that intact reactivity portends less future depression.[21]

In the most recent, and in some ways most convincing, study so far — conducted with colleagues in the Netherlands — we examined whether these relationships held when reactivity was assessed outside the laboratory in everyday life. Forty-six depressed outpatients from a community mental health center in Maastricht, a medium-sized Dutch city, enrolled in the study. Their emotional experience was repeatedly sampled as they went about their daily routines for six days. The sampling was done via a small wristwatch that beeped to prompt the person to report on feelings and any positive or negative events that had occurred since the last beep. After this sampling of emotional experience and daily events, the patients all received a course of pharmacotherapy (antidepressant drugs) alongside supportive psychotherapy. Finally, all were followed clinically for eighteen months to ascertain who remitted and who did not. During the eighteen-month period, twenty-six patients achieved remission from their depression and twenty did not. Remarkably, as we saw previously, it was those patients who had *less* negative emotional reactivity to the daily negative life events during the week of sampling who proved to be less likely to recover from depression over the next year and a half.[22] A depressed person who still had the capacity to get bent out of shape was more likely to recover.

––––––––––

SOMETIMES PEOPLE USE the phrase *nervous breakdown* in connection with depression, as if it were a matter of a machine falling into disrepair, a junk pile. But the-harder-the-shutdown, the-worse-the depression pattern suggests a more troubling reality. Rather than turning us into inoperable junk, depression actively shapes our thoughts and behaviors in specific ways that keep us depressed. In fact, deep depression is so good at breeding depression that it often extends many months or years beyond any instigating crises. This is one of the most curious things about deep depression: how often it outlasts its apparent source.

There are several possible explanations for this. One is that a lag between environmental improvement and mood improvement is built into depression's advance directive. Let's return to our thirty-six brave volunteers from the Minnesota Starvation Experiment. For weeks the men had looked forward to the extra rations they would get when the deprivation phase of the study ended and the refeeding stage started. Yet even six weeks into the refeeding stage of the experiment, the mood of the thirty-two remaining men continued to drag; morale lifted only slightly at the news that Japan was nearing surrender in the war.[23] This lag in the mood system may seem cruel, but it may be part of depression's design, to hold behavior in place until depleted resources can be rebuilt. Consistent with this logic, it took weeks after refeeding for the men to begin putting on weight, even with greatly improved rations. Eventually the men became themselves. Writ large, we should rarely expect good fortune, even very good fortune, to register immediately on a serious depression.

A second possibility is that a biological response strong enough to shut down seeking may have unintended consequences. Let's not forget what a great feat it is to shut down behavior. Seeking is our natural state. It's not only that we work hard for big payoffs like a first kiss or major promotion; we also seek in neutral situations when no rewards are close at hand. Like cars, we creep forward at idle. Consistent with the idea that the default settings of the mood system are slightly positive, most people report more positive than negative affect when they are in a neutral situation (e.g., at the start of a research study). Psychologists label this phenomenon *positivity offset*. One obvious reason for positivity offset is that it encourages exploration [24] and engagement with novel objects. Historically, to find a delicious new berry or sheltered campsite, a bit of wandering has been good for fitness.

Shutting down seeking requires a potent mechanism that is likely to bring with it collateral damage, including the possibility that deep depression is sustained well beyond an instigating crisis. Depression is sluggish by design, so it's not easy to draw a bright line where the designed response ends and the collateral damage begins. Still, it's unlikely that every effect of depression represents design. Two examples of bad physical health outcomes increased by depression are risk for high blood pressure and risk for heart attacks. These outcomes are so harmful and utterly without benefit, it's impossible to conceive of them in terms of evolutionary design.

The idea of collateral harms is attractive even if the exact mechanisms that create them are currently elusive.[25] After all, the body in depression undergoes many changes; it is a challenge to identify which might unwittingly sustain episodes of depression. At this point several suspects have emerged. One is elevated cortisol, already mentioned as a stress hormone important in bodily mobilization for use in short-term emergencies, such as a zebra needing to escape from a lion. Normally, cortisol is tightly regulated to help it return to low levels, but this hormone is persistently elevated in many depressed people. Longer exposure to high cortisol has a welter of physical effects, like wasting muscles and even damaged neurons in the brain. Prolonged cortisol exposure may also help stretch out depression. Consistent with this notion, patients with Cushing's syndrome, a metabolic disorder caused by high levels of cortisol, often become depressed. Medications that block the effects of cortisol to treat depression have shown some promise and are in an experimental phase of testing.

Another suspect group of substances is the inflammatory markers, such as cytokines. These chemicals foster the inflammatory response, classically seen in swelling at the site of injuries or infection, which ultimately helps repair tissue and heal wounds. Like cortisol, the proinflammatory cytokines are high in depression. We don't fully understand why this is the case. Nor do we understand all of their effects in the body. Nevertheless, there is evidence that when a person has high levels of cytokines, he or she will report feelings of sickness and fatigue. The more cytokines, the stronger the feeling of malaise. Moreover, in studies in which cytokines are administered acutely, receipt of these substances produces depression-like symptoms.[26] These substances, too, are a plausible culprit for collateral harm.

It is important to remember that although evolution has powerfully shaped our capacities for mood, we are not its prisoners. Even with a powerful evolutionary directive to become depressed, we retain a margin of control to shape its course. There are things people can do both to make depression worse and—as we will see in the next chapters—to make it better. Indeed, the bad choices humans sometimes make are another major reason that deep depressions stretch beyond their instigating crises.

Take a depressed insurance salesman who has fallen behind on his sales targets for three straight months. After receiving a stinging performance evaluation, he impulsively walks to the corner office and tells the boss to shove it, getting himself fired. Now in addition to mood problems, the man has economic problems. Life stress is not always a bolt from the blue. Psychologist Connie Hammen has called this process *stress generation*, a term that reflects that depressed people can exacerbate depression through the consequences of their own behavior.[27] One troublesome behavioral response, repeated all too often, is self-medication of mood through heavy consumption of drugs and alcohol. Perhaps one in five depressed people abuses drugs and alcohol, a strategy that may provide relief in the very short run but cements depression in the long run. Ann Landers was right: "When alcohol wears off, you will be more depressed than ever."[28]

A Look Forward

The core of depression is a slow-moving black hole for motivation. There are probably multiple reasons for its slow motion: because lags are built into depression's basic architecture, because shutting down behavior likely creates collateral damage, and because humans make problematic choices that extend it in time. Taking them all together, we can start to understand why deep depression is so trenchant and often impervious to psychological and psychopharmacological treatment. Anyone who promises an easy solution to depression does not understand what we are up against. However, despite the consuming and self-perpetuating nature of deep depression, people do emerge from the black hole, to partial and sometimes even the glory of a full recovery. In the following chapters we consider what enables these successful escapes.

CHAPTER 8

An Up and Down Thing: Improvement in Depression

WE KNOW QUITE A BIT ABOUT WHY AND HOW PEOPLE FALL INTO depression. From negative thinking to stressful environments to poor ways of coping to gloomy temperaments, falling in is no great mystery. Yet if you're in a serious grinding depression, what you yearn most for is improvement—for a sign, any sign, that today might be even marginally better than yesterday. Deep pessimism is engaged in a pitched battle with yearning. Researchers know far less about how to get out, even though there are numerous examples of people who have successfully done it. How do people pull out of depression?

I certainly wanted to believe that improvement was possible. Eighteen months into my depression, I walked on Woodlawn Avenue, on the outskirts of Baltimore, a willful act to keep moving. Was this walk better than the last walk, or the one before that, or the couple hundred walks of desperation before that? Even though I wanted to believe it was, it did not seem any better. Traffic, and the whole world, buzzed by me. I still could not think straight. I felt completely obliterated as a human being. My thoughts were as desperate as usual. When I noticed other people, I wondered what it was like to be alive. They did not know, could not know, how I felt inside. My shell still passed for normal. I felt like I should scream for help, someone should help, but I knew that the time for screaming had passed. Best to just keep on walking, walking dead, one of the few things I could still do. So I kept on walking. Optimism spent, still I couldn't accept an unacceptable fate.

A consoling fact is that most people do pull out of depression. In one careful observational study, Martin Keller and his colleagues followed a cohort of 431 patients diagnosed with depression—many of them so debilitated that they had been hospitalized—over a five-year period. Two months into the observation, nearly one in three had recovered from the episode. By six months, over half the patients had recovered. By five years into the follow-up, 88 percent had recovered.[1] These were individuals suffering from the most severe depression, so we could hypothesize that less-affected depressed persons would recover as or more quickly. Data from shorter follow-up studies of outpatients generally show remission rates similar to or better than those Keller found. Likewise, data from samples that are more representative of the average depressed person in the community suggest that depression will last a year or less 90 percent of the time.[2]

Experts have historically referred to depression as a self-limited condition, a problem that ends by itself.[3] The notion of self-limiting is familiar from biology: the growth of a colony of organisms may be self-limited because once the colony gets to a certain size, either its waste products become toxic, or the colony starves because all available food has been consumed. Applied to medical diseases like the common cold, self-limiting means the body's immune response is adequate to keep the cold virus in check. We can't specify the details of the self-limiting process that ends depression, so it is tempting to switch to metaphor instead.[4] Depression is like a forest fire that burns out when all the woodland fuel is spent.

Patients are almost always told that depression is self-limited in order to help them believe that they will get better. This is a therapeutic bromide conveyed with the best intentions. As long as a patient doesn't ask too many questions, it provides comfort, at least for a time. But as depression grinds on, vague bromides don't work as well. For someone at the end of her rope, whose patience is measured in days or hours, it's the pace of improvement that is critical. Someone who has already had the best years of life torn up by depression wants to know, "*When* will I be better?" Hearing that most people recover eventually, even if it is true, is not good enough.

The second basic fact about improvement in depression is that it can be accelerated by formal treatments. There are many treatments available; for three specific treatments there are systematic data to support claims of effectiveness: antidepressant medications, cognitive-behavioral therapy, and interpersonal therapy. However, like the notion of a self-limiting illness, the existence of supported treatments is not quite as comforting as it might at first appear. Yes, these treatments accelerate the recovery process, on average, but the result for any given individual is unknowable in advance. One of the biggest frustrations for professionals and their patients alike is being unable to predict with confidence which treatments are likely to work. Despite efforts to identify shortcuts that will predict who will respond to what, making a treatment recommendation remains largely a matter of guesswork. Furthermore, we don't know in advance that *any* treatment, even one of the proven ones, will be effective. We know only that trying a supported treatment is a better bet than not trying one at all.

That we know less about how people climb out of depression in part reflects the influence of the disease/defect model, which is oriented toward identifying the factors that produce depression rather than those that obviate it. Finally, our knowledge of improvement remains limited because studying improvement is difficult, practically speaking. It requires identifying a sample of depressed people and following them carefully, repeatedly, and over many years, a process that requires large investments of time, labor, and money. One shortcut is to study the course of depression in people who have sought treatment for it; in fact, much of what we know about improvement and what predicts it comes from studies of treatment-seekers. These are often studies done over the short term that address the question of why a particular treatment works (or doesn't work), rather than why depressed people in general improve. Our knowledge of how improvement plays out in real-world settings, where people might not receive treatment or might frequently switch treatments, is relatively slim. Nonetheless, data on improvement, albeit imperfect, arm us with something more than the basic fact that recovery from depression is a matter of time, shortened by supported treatments.

The overall story is a tale of two trajectories of improvement. For about one-third of depressed people, improvement is relatively rapid and is often, but certainly not always, sustained. For the other two-thirds, the trajectory is flatter. Improvement often stalls out, and gains are painfully halting, uneven, and fragile. The sickening possibility of falling back into depression is ever present: two-thirds of those who partially recover will fall back into depression. Relapse—deep depression returning before it ever fully leaves—is perhaps the cruelest outcome of all.

Why do some people spring forward and others stall out and fall back into a depression that becomes chronic? Mood science helps make sense of this often zigzag path of improvement in depression.

Springing Forward

We would all want to be in the lucky third. In study after study, about one in three depressed people improves early and tends to retain and expand upon his or her gains. We see the segment of early gainers across the spectrum, in the mildly depressed as well as in severely depressed people.[5] In one of the more convincing studies, Armin Szegedi and colleagues tracked more than six thousand patients who had been enrolled in antidepressant drug trials. They concluded that marked improvement early on, in the first two weeks, was an extremely strong predictor of patients' ever improving.[6] Another estimate states that about 60 percent of the improvement occurring on antidepressants in drug trials happens during the first two weeks of treatment.[7] This quick pace challenges an older conventional wisdom in

psychopharmacology that it takes many weeks on antidepressants for action in the brain to manifest clinically.[8]

The early gainers show up like unexpected houseguests
throughout the literature on the course of depression. In research on cognitive therapy, scientists have been perplexed
by what animates the "sudden gains" shown by a subset of
patients who undergo this treatment.[9] More curiously, early
gainers show up not only among people taking medications
in drug trials but also among those on placebos.[10] And early
gains have been found in groups of depressed people who are
residing in the community and aren't involved in any formal
treatment.[11] Further, nearly one in five depressed people on a
waiting list for treatment demonstrates improvement during
the waiting period significant enough to ordinarily be considered a "treatment response."[12]

These early gainers are of keen interest, not only because
they get relief quicker but also because their improvements
hold up fairly well. Early improvement can be an important
stepping stone to a fuller and more enduring recovery. People who improve early appear to have better long-term outcomes than those who have more gradual improvement.[13] In
psychotherapy process research, there has been considerable
interest in what might explain sudden gains and what these
gains signify.[14]

What are the early gainers telling us? One hypothesis is
that "treatment works." Although this interpretation is intuitive, it's not clear that miraculous treatment responses are the
sole or best explanation here. They wouldn't explain why
there is the same segment of early improvers among patients
who take placebo pills, patients on waiting lists, and peo-

ple in the community who are not in treatment of any kind. Starting therapy or taking medication is not required to experience early improvement.[15] When rapid gains occur at the start of a treatment, they may be coincidental and not a product of the treatment at all.

The commonness of early improvement suggests that it may not need a special explanation; it may be part of depression's natural history. Yet understanding why some people experience a more benign course of depression has tantalizing implications for how improvement works and for how episodes of depression might be shortened. Although surprisingly little is known about the characteristics of early improvers, we can formulate some educated hunches about what might be going on, based on the general outcome literature and our knowledge of what drives the mood system.[16] Let's take up each of these hunches in turn.

Hunch 1: Early Improvers Face Fewer Complex Life Problems

Recall the predicament of the mood system. Its task is to sum up what the environment holds, its available threats, and opportunities for action, and to direct behavior forward or allow us to stay in place. Consider Jane, a young woman who is depressed in the wake of the acrimonious breakup of an intense romantic relationship. This kind of discrete problem might benefit from a cocoon of low mood—a protected pause to analyze the past and develop solutions. Her low mood is likely to pass as she makes decisions about the future, whether that means searching for a man with different qualities than her ex or taking a break from dating entirely. Jane's story typifies one sort of early improver, someone whose depression enabled her to arrive at a reasonable solution to a vexing, but manageable, problem.

What about when multiple problems arrive simultaneously? Imagine if on top of the breakup, Jane's mother was diagnosed with Alzheimer's. Increasingly frail, her mother can no longer live independently. Jane's younger brother has decided he can't cope with the situation, leaving it to Jane to arrange all the medical care and housing. In this scenario, with multiple open problems simultaneously, Jane is less likely to be an early improver.

Depressions, of course, aren't always tied to a specific life event; they often come attached to a variety of slow-moving difficulties, whether psychiatric, medical, or psychosocial. Notably, these are problems like severe symptoms of anxiety, substance abuse, chronic pain, and the burden of serious ongoing health issues.[17] And new stressors can result from the depression itself. If a depressed man can't concentrate, his job performance suffers as a result, which leads to his losing his job when his company downsizes, meaning additional financial difficulties.[18] Consistent with the idea that combinations of problems may be less manageable than individual problems, one study found that having three stressful life events was actually four times as bad for future depression as having two stressful life events.[19] With each new problem, the mood system must face a longer and more complex equation with more unknown terms. When there are no solutions at hand, we should expect sluggish mood improvement at best.

Hunch 2: Early Improvers Have
Secret Weapons Against Depression

Based on what we generally know to be helpful in bringing depression to heel, we can surmise that early improvers try to exercise, keep active, and maintain a normal schedule. If they are in treatment, they keep their appointments and follow the treatment plan. They try to make full use of their social support networks rather than immediately withdrawing from them.

Early improvers may also benefit from an innate resilience that helps them to instinctively do the right things. Nicole Geschwind and colleagues found intriguing nimbleness in the mood system of early improvers. These investigators gave pagers to a group of depression patients, then prompted them to report on their emotions several times a day as they started a drug treatment. Some patients demonstrated early improvement in positive mood states, during the first week of treatment. Mind you, these patients still had many of the symptoms of depression: they couldn't concentrate, felt pervasive guilt, still had problems sleeping, and continued to be burdened by a sad mood. Yet there was a subtle shift early for some patients who experienced glimmers of enthusiasm and cheer. Such early shifts in positive feelings foretold that a patient would achieve remission by the sixth week on the drug.

The difference was remarkable. Those whose mood shifted quickly were in fact thirty-four times more likely to achieve eventual remission than patients who did not experience the early shift in feelings.[20] How did a rapid change in feelings bring about the end of depression? One possibility is that some individuals' brains were unusually receptive to the pharmacological effects of the drug, and for them the change toward more positive feelings represents an early drug response. Another, perhaps more interesting, hypothesis is that for some patients even a small experience of positive feelings is enough to catalyze behavior, allowing them to reengage with their environment, seek out rewards, and benefit when good things happen.

Indeed, one reason to intensively study the early improvers is to learn whether they employ any tricks or secret weapons that may help account for their success. This work can't start soon enough.

Hunch 3: Early Improvers Are Lucky

If bad fortune blocks improvement, a lucky turn at the start of a depression might be enough to stop it in its tracks. Although there has been little study of positive life events and early improvement, the occurrence of positive events often predicts improvement in general. A group of Dutch investigators found that positive life changes increased in number during the three months prior to recovery in primary care patients with depression and anxiety.[21] George Brown and Tirril Harris, life stress researchers in England, have focused on a class of events that they call *fresh start* events. These events inject hope into a situation of ongoing deprivation, such as when a financially stressed young, single mother meets an economically stable man and they decide to start a household together. Brown and Harris have shown that when fresh-start events occur, they can bend the course of an entrenched depression in an upward direction.[22] Early improvement can reflect improved life circumstances.

One thing that intrigued me as I read the positive life events literature was that the effects weren't stronger. In a number of studies, the good events tallied up by the investigators had no decisive effect on depression.[23] From the perspective I have laid out, this is puzzling: if the mood system is the great integrator, it should be attuned to all of our progress toward all the major goals handed down by evolution. As we do better with respect to survival, reproduction, status, or allies, the depression should invariably dial down. How could it be that a depressed person's improvement after positive life events in many cases is not dramatic, or even noticeable?

This puzzle hits me at the most personal level. On December 5, 1995, a beautiful baby girl named Sophie came into the world. Minutes after her delivery, I held her and peered at her lively features. What could be clearer evidence of your survival and reproduction than to gaze into the alert eyes of your baby daughter? Yet my mood system would have nothing of it. Everyone congratulated me on the baby and told me how happy and proud I must be. It was true, she was rather a good one. Still, as they handed her to me, my dominant thought was *I am just not ready for this*. And I wasn't.

I was just as miserable as before. The situation was a strange one: I had a wife who unaccountably refused to give up on me, and a beautiful daughter, and yet with all of this obvious purpose in life, I remained hideously and fixedly depressed. Sophie was conceived. I tried another antidepressant. Sophie was born. I tried another antidepressant. Sophie scooted across the floor. I was as big a wreck as ever. My

mood simply did not seem interested in the ebbs or flows
of my life, which I took as final confirmation that I was
irretrievably gone. My long rambles continued; now a baby
rode on my back.

The operations of the mood system are not transparent.
As it integrates the state of our inside and outside environments, the system does not reveal its logic or priorities to us.
Why did Sophie's arrival initially make no difference? Was
it the overwhelming psychobiological momentum of a depression, which for a time steamrolled everything in its path?
Was my mood system making a computation that this was
an unpropitious time for a baby, arriving into a strained marriage, to a father who thought his mind had left, and in a time
of grave uncertainty, economic and otherwise? Or was my
mood system simply rebelling against any further revision to
an established life plan? This puzzle is difficult. We cannot
be sure.

What was certain is that my life plan had no entry for
stay-at-home dad. This was an improvisation. Even if, in-
tellectually, I knew that this was an opportunity, a chance
to grow, my body did not warm to the task. I felt viscerally
incapable of being a father. My best efforts covered the law:
diapers were changed; I could rally when Sophie cried. But
what could I do well? I was good at lying down. Lying down
hurt the least. I liked it when Sophie napped; I could lie down
then. I feared my girl would soon figure out the truth, that
she had a broken dad. She saw me every day, and yet I could
not let her see me this way. The urge to run away and hide
was great. I needed to get my act together rather urgently.

Had I been in a research study during initial fatherhood,
my data point would have counted against the hypothesis

FIGURE 8.1. With Three-and-a-Half-Year-Old Sophie in June 1999, After I Had Substantially Improved.

Photo credit: Robert Rottenberg

that the mood system takes full measure of positive events. Thankfully, my data subsequently became more ambiguous. By the time Sophie was two and about to begin preschool, I was finally improved, and substantially so (see Figure 8.1). Something had broken the momentum of depression. But what? Was my mood system registering the success in parenting an ebullient toddler, keeping my family intact, and becoming reengaged in a new career possibility, or was some other hidden, self-limiting process at work? Or did all of these things combine in some synergistic way?

And what about my lovely Sophie: Did she provide the critical jump-start? Thankfully we cannot go back and subtract
her from the equation to find out. It would be hard to know
where to look for a jump-start; there was no single moment
when the corner was turned. She took her first steps, said
her first word ("bird") on one of our rambles, and wore the
king's crown at her first birthday party. My mood system
eventually started to believe in the future. I started to claw
my way out of the depression. The progress was so slow,
the change was unnoticeable day by day, yet over months
and years the impossible—improvement from a devastating
depression—happened.[24]

Grinding Onward

My experience of slow and halting improvement is unfortunately more the norm than the exception. Being mystified by
the process of improvement is also not atypical. Sufferers are
often humbled by depression's meandering course. Scientists
also have cause for humility (even if there is no glory in admitting the limits of what they know). Part of why we are
not winning the fight against depression is that we lack a
comprehensive understanding of how improvement works.
We have only clues.

There are many shades of improvement between deep
depression and full wellness. There is no obvious point to
denote where a disruptive clinical mood state ends and a
"normal" mood state begins. Yet as depressions go from black
to gray, each shade of improvement matters. Each stepwise
decrease in depressive symptoms brings with it a stepwise im-

provement in functioning. Large samples that follow the same people over time show that with each downtick in symptoms, a person will be more able to work, more able to go to school, more able to be a good spouse, and so forth.[25] While the researchers can demonstrate the impact of even small gradations of depression, sufferers may not notice small changes, especially if they occur slowly. Deep depression's catastrophic implosion can fade imperceptibly into a merely horrible mood, which can fade imperceptibly into a garden-variety bad mood, which can fade imperceptibly into complete normalcy. With improvement so gradual, there is no distinct moment when depression ends; it has no "last day."

We also know that the sequence of improvement, the order in which people lose symptoms, is lawful. Brian Iacoviello and his colleagues found evidence for the phenomenon of symptom rollback.[26] People who experienced depression were followed across several episodes, which allowed the investigators to date the precise order in which symptoms were acquired and lost. This order was surprisingly regular; symptoms were generally lost in reverse order to that in which they were originally acquired. The first off were the last on, and vice versa. As depression remits, it can repeat, in reverse order, previous stages and experiences. There is even parallelism in the length of these stages. People who have a longer improvement period tend to have had a longer period in which depression symptoms first mounted. In other words, the slower the wind up, the slower the wind down. Within this orderliness, there is also variation. Because different people acquire symptoms in different orders, what symptoms predominate during the improvement period varies radically.

For one person that stubborn, lingering symptom may be insomnia, for another it may be guilt, for still another it might be fatigue. Different people are shaking off different things.[27]

For Sylvie, the strangest and most distressing symptoms were the first to go. First she noticed she could move. She was still in the hospital a week after her suicide attempt, "afraid I would never get well." For weeks she had been stricken with leaden paralysis; it hurt to open the mailbox. Now she noticed gradually that she could move without being in agony. The next thing that came back was the ability to eat. For months during her depression she hadn't been able to swallow food. She loved food and tried to eat but, "after 2 or 3 bites my throat would close off." She lost twenty-five pounds. Now she was eating again, without any relish, but to be able to eat at all was itself a pleasure. Sylvie was guarded about the meaning of these signs as she went back to work. After three or four months of improvements were sustained, she could allow herself to be hopeful that she would get better.

The last symptom to go was sleep disturbance. Consistent with the rollback phenomenon, this was also Sylvie's first symptom to come on. As her depression escalated, she would lie in bed night after night, lacerating herself for her failures as a mother, fearing for her daughter's future, a fury of despairing thoughts. Now she was more patient with herself and nurturing, not critical. She listened to and benefited from meditation CDs. She reminded herself that she could take Trazodone (an antidepressant that is also sleep inducing) if she needed to. "What's the worst thing that happens? You take the Trazodone." As she lay awake, she reminded herself that she was beating the depression; she was winning her life back. She had survived, and now she was going to be better than ever before. Eventually Sylvie fell asleep. With time, even this last symptom disappeared.

Improvement generally is a matter of time, but the longer a depression has gone on, the more time it seems to take. Martin Keller and his colleagues documented this by following patients over many monthly intervals for many years. If a patient had already been depressed for three years, there was only a 1 or at most 2 percent chance that person would start to recover in the next monthly interval.[28] This is the disturbing flip side of the phenomenon of early improvement: if you miss the early improvement boat, recovery may take a long time.

In a fraction of depressions, improvement is arrested; when depression stretches out for at least two years, it may be designated chronic.[29] What arrests improvement will come as no surprise. As one commentator put it, "The determinants of chronic depression do not differ qualitatively from acute depression."[30] In other words, everything we have talked about thus far that breeds low mood and depression—a neurotic temperament, a mood-punishing routine, or exposure to trauma early in life—also breeds chronic depression. People with chronic depression simply have more of these factors than people who have shorter depression episodes.

This also applies for life stress. Just as a stressful life can make you depressed, continuing exposure to stressors maintains depression. It is a major factor blocking improvement. For example, in treatment studies people who had discordant and conflictual relationships with their significant others responded less well to three different depression treatments over three months than people who were in more harmonious relationships.[31] Another treatment study made a detailed assessment of each person's social circumstances at the outset of the research and again twelve weeks later, after randomization to a treatment. Patients who faced significant environmental adversity at any point were about half as likely to respond to the treatments and remit from depression as patients who were in more benign environments. Interestingly, not all aversive events were equal; negative events involving a theme of humiliation or entrapment, such as a woman with crippling arthritis living with a highly critical and sometimes violent partner, interfered with improvement most directly. Aversive social contexts blocked improvement regardless of the kinds of treatment the patients received, with less than one-fifth of those in aversive social contexts remitting after twelve weeks of treatment. Though most treatment studies assess only the short run, effects of severe life events have been shown to retard treatment response even forty-five weeks out, evidence that they contribute to chronic depression.[32]

Yet although similar factors propel the onset and the continuation of depression, it would be wrong to say that chronic depressions are identical to briefer depressions. When depressions become chronic, they impose at least three special burdens. Simply by virtue of their length, they are inordinately taxing. As Virginia Heffernan put it, "Unless you are rich, and can convalesce in a sanatorium estate (where visitors come down a tiered, oceanside lawn to find you at your easel), you have to keep going when you are depressed." For an ordinary person, "That means phone calls, appointments, errands, holidays, family, friends, and colleagues. For me, this is where things got tangled. Depression brought to me a new rationing of resources: for every twenty-four hours I got about three, then two, then *one* hour worth of life reserves — personality, conversation, motion. I had to be frugal while I was hustling through a day, because when I ran out of reserves, I lost control of what I said."[33]

Chronic depressions more profoundly interrupt and disturb life plans than do shorter ones. Think of that person with the hole in his resume. That student who went from As and Bs to dropping out of college. And chronic depression has greater power to alter a person's self-concept than briefer episodes do. Depression for months and years can potentially harden thoughts of the self as fundamentally worthless, powerless, and ineffectual. As depression drags on, autobiographies are rewritten. Depressed people become unable to remember happy times, or times when they even had a normal mood. The very concept of a normal mood itself becomes alien. I've had more than one depressed participant offer me a pained smile during an interview when I asked, "When was the last time you felt like your usual self?"[34] These are the disorientations of chronic depression.

Toward Limbo

William Styron, in *Darkness Visible*, his haunting evocation of depression, pointed out that ordinary language often fails to describe the experience of depression rising and falling. He cited Dante's *Inferno* as a faithful representation of "this fathomless ordeal." Styron commented, "For those who have dwelled in depression's dark wood, and known its inexplicable agony, their return from the abyss is not unlike the ascent of the poet, trudging upward and outward out of hell's black depths and at last emerging into what he saw as 'the shining world.'"[35]

Sooner or later, with enough fortitude, a person with depression does emerge from the darkness. But the journey is not complete. For some, this improvement will be the beginning of a sustained recovery in which the world will indeed shine. However for many, if not most, subtle signs of depression will lurk and linger. Standard treatment of mood disorders, even in specialized settings, tends to leave residual symptoms. The patient is almost fully well, but not quite.

Is almost well good enough? What forms of suffering we find acceptable is a moral and social question rather than a scientific one. Our culture provides voluminous feedback on what a person should feel and what symptoms should not be tolerated. My mood science perspective takes a neutral stance. No point on the spectrum of mood is defined ahead of time as the normal, preferred mood state. We have a range of moods and mood symptoms appropriate to the state of our internal and external circumstances. Residual low mood, like deep depression itself, is a natural product of our mood system that involves both costs and benefits.

That said, although evolution constrains us, we are not its prisoners. Just because we are not wired for bliss, we need not acquiesce in low mood. Actuarially, it would be unwise to acquiesce. We know from epidemiology that residual depression symptoms are one of the strongest predictors of the return of deep depression. In this respect, even when depression is nearly vanquished, the stakes remain high. In the final chapters I draw on the mood science perspective to think through what we mean by recovery. I consider why the mood system drags along with residual low mood even after an episode has largely passed, as well as why these limbo states often put the sufferer in a precarious position.

CHAPTER 9

In Limbo

Whatever that monster is that's been on your
shoulder for three years, now it's off. But you're
still jumpy. You are wondering how much stress
you can handle. You feel leery and shell-shocked.
You don't see how you could ever return.[1]

Cliff Richey

BOOKS ABOUT DEPRESSION TYPICALLY DO NOT INCLUDE A CHAPTER
about when people are nearly better—what comes between
the worst passing and the achievement of wellness. Severe
self-loathing is replaced with vague dissatisfaction, sleep is
better but still sometimes fitful, and in the place of plunging
low mood is a dull emptiness. However, those who experience residual depression know firsthand that it can be depression's most awkward—as well as its most precarious—phase.
It is a state of limbo many times over, one that may determine

the course of recovery. Because the disease model of depression trains its sights on full-scale depression, residual depression has by comparison received only residual interest. Even the phrase *residual depression* will be unfamiliar to many. Depression in its twilight is left in a kind of conceptual limbo.

Being partially back in your life is confusing. Uncertain whether depression is really and truly gone, you have no script for how to act. As author and journalist Tracy Thompson put it, "I yearned to get better; I told myself I *was* getting better. In fact, the depression was still there, like a powerful undertow. Sometimes it grabbed me, yanked me under; other times, I swam free."[2] After a hospitalization for depression, she returned to her job at the *Washington Post*, but cognitively she could only focus intermittently. "I wrote long memos to remind myself of what I was doing, sometimes only to find a similar memo I'd written several days before."

In addition, there is a social limbo. One of the most powerful impulses during a deep depression is to retreat from others, to withdraw. As depression improves, the sufferer begins to reengage, to answer the phone, to make plans, to be emotionally available to others. This can be a time to assess the damage done to relationships, to send out feelers. But gains are tentative. After the worst has passed, neither the sufferer nor her social partners know where they stand. What are the rules now? It is relatively easy to make allowances for a depressed person who is incapacitated, in the throes of an episode. But what allowances should be made for someone who is mostly better?

During my own limbo, I still had black periods when I wanted to lie down. Sometimes I gave in. I remember my wife Laura asking me, "Do you want me to treat you like a sick person or a well person?" I paused because I didn't know. Spouses, bosses, and friends who were sympathetic throughout disclosure and treatment may lose patience with making allowances. So, too, the sufferer may be done with depression, well before depression is done with him. The price of saying "all better!" prematurely is bearing the weight of lingering depression alone, without proper support. After the worst is over, there is a powerful incentive for all to pretend that everything is back to normal. Limbo is too complicated to explain.

This was true of Cindy Richardson, a therapist in Milwaukee, a few weeks after she had climbed out of her most recent depression. Cindy has experienced episodes of depression on and off for the last twenty-four years. At the tail end of her most recent episode she felt "both the possibility and the presence of a little depression." In this guarded period, she reported feeling "a little crazy," "unpredictable," "self-conscious," and "inconsistent." She worried that "people will think I'm unstable, or unreliable." Despite knowing as a therapist that depression does not reflect a personal weakness and despite her knowing the value of seeking social support, she kept her lingering pain mostly to herself, even hiding it from her significant other, Dave. Cindy tried to be "self-reliant." She didn't want to "be the center of attention when she's down," for fear that friends would pull away from her. She longed to be well again, to be "almost forgetting about depression."

People with residual symptoms are also in limbo when they seek help. Most of the training that practitioners receive, and most formal depression treatments, are aimed at acute depression. For professionals, the residual phase of depression is a fog zone.[3] There is no consensus on how aggressively it should be treated or for how long. Given finite resources, treatment development naturally focuses on the most dramatic and distressing symptoms of depression's acute phase. Still, it's telling here that clinical trials conventionally measure success in terms of reducing depression symptoms, not eliminating them.[4] The yardstick of success in treating depression is to achieve 50 percent or greater symptom reduction. Another yardstick is remission, defined as when depression falls below a standardized symptom cutoff, most often below a 7 on a symptom measure called the Hamilton Depression Inventory.

There's no question that it is good news when a person reaches these milestones. Yet it's eye opening to consider how much depression can remain even when such levels of improvement have been attained.[5] In one representative study, 80 percent of people who were treated with Prozac for eight weeks and who met the study criteria for full remission were left with low energy, concentration problems, or other symptoms of residual depression.[6]

Success should mean that depression has been completely treated and that any remaining symptoms are of no real concern. Yet recent data have shown that over the long haul, small amounts of residual depression have surprisingly bad effects. Eugene Paykel and his colleagues found that among those who had improved substantially, residual symptoms tripled the risk for a subsequent return to deep depression. Relapse occurred in 76 percent of those with residual symptoms and only 25 percent of those without them.[7] Similarly, a large cohort study found that people who had residual symptoms at recovery relapsed much faster—on average over three times faster—than those who were asymptomatic at recovery.[8] These same patterns hold when people with a first lifetime episode are followed for up to a dozen years. People who have residual symptoms during recovery have more severe and chronic future courses, with deep depression recurring sooner and more often.[9] In other words, "almost better" is worse than you might think. We need sounder strategies for addressing residual depression.[10] One reason we're not winning the fight against depression is that our available treatments leave so many in partial recovery limbo.

Sara Matthews, a single, unemployed Memphis woman, is one of the millions of people inhabiting the in-between world of partial recovery. She has battled depression on and off for fifteen years. Of late she is doing better, helped by a new combination of three medications. When her therapist had her complete a depression questionnaire at a recent session, Sara scored as mildly depressed, which didn't surprise her. Yet her lingering depression complicates her job search. "You know what you have to do, you just can't do it," Sara says wearily. "It's like you have bricks on your feet." Her indecision makes it difficult to focus in her search, and lingering impulses toward self-criticism sap her ability to convincingly sell herself to employers. In the past she has settled for "crap jobs," and low mood erodes her confidence that she should ever aim higher. Her treatments, all pharmaceutical, have been juggled dozens of times over the years, but she has not been able to put together a sustained period of back to normal. She is consoled only by small patches of feeling good that last a week here, a week there. Sara believes that if she can put low mood behind her, a more extroverted, fun-loving side of her will come to the fore. A natural storyteller, she talks about the parts of her personality that have been submerged by depression. There's a lift in her voice. Sara is impatient for improvement. Sometimes, "I want to just jump out of my skin and be free."

Many share Sara's story. But because residual depression falls between the cracks, it is understudied. From a mood science perspective, it is just another mood state, a part of the natural history of depression that we need to document. Limbo is the shallow depression that follows a deep depression. The reasons that limbo persists overlap with the reasons that low mood persists, a theme I took up earlier.

Unresolved stressors can put a ceiling on mood improvement. These can be the fallout from deep depression—a marriage strained, a career or education disrupted—or they can be new. In Sara's case, concern about underemployment is a persistent theme, one that has shadowed her mood problems as they have waxed and waned over the years. Sara has long felt uncertain what kind of work she could or should do. After changing fields several times, she is thirty-eight. She regrets not capitalizing on an earlier opportunity to deejay on a radio show. Underemployment saps her recovery; she worries her job search will drag on.

Even when mood improves, progress may hit a ceiling if daily routines continue to be depleting, limiting how much improvement is possible. Sara says she knows that sunlight and exercise are "free antidepressants," a statement that is borne out in mood research. But she doesn't act on this knowledge. Instead, she often pads around in her pajamas until midafternoon. She persists in following a routine she knows to be mood depleting.

A person's temperament, her inborn characteristic responses to stress, may also limit mood improvement after deep depression. Sara knows full well she is prone to brooding—depression runs in her family. For years she watched her mother struggle with several punishing episodes of depression. Sara's psychiatrist suggested she start cognitive-behavioral therapy to break this intergenerational cycle, to try to chip away at her tendency to dwell on the negative.

Our best data show that about two-thirds of those who recover only partially will fall back into an episode of deep depression. Like shallow depression, limbo is dangerous. During this precarious stage, mood-worsening behaviors, like remaining committed to unrealistic goals, engaging in destructive responses to low mood (e.g., excessive drinking, drug abuse), and retaining unrealistic expectations for mood, afford routes through which deep depression can reemerge.

In addition, with relapse there is something new and frightening. We have strong evidence that the mood system has an easier time going from limbo *back* to deep depression than it did getting there the first time. Less adversity is required to push a person back down. A rich vein of studies of life stress and depression consistently shows that second and third episodes of depression aren't as closely linked to severe life stressors as the first episodes are.[11] These findings suggest that one bout of deep depression can change how the mood system behaves going forward.

Proponents of the disease model see depression's tendency to recur in terms of a pathological breakdown; those who get depressed again become ever more defective. Psychiatrist Robert Post has proposed an influential biological model to explain the ease with which depression recurs, which he calls the "kindling model." This model analogizes the progression of mood episodes in people to the course of seizures in other animal species. A key observation is that kindled, or artificially induced, seizures can progressively change the brain, such that less electric current is required to generate additional seizures.[12] Over time less and less external provocation is required, until the seizures eventually become spontaneous. Applied to depression, the idea of kindling is that episodes of depression likewise progressively change the sufferer's brain; with each episode, less stress will be required from the environment to make depression recur.[13]

There is no question that depression changes the sufferer in more ways than one. And it is plausible that the changes wrought by depression ease the path for its return. Still, the metaphor of kindling—with its implications of a pathological process and damaged brains—is problematic—and probably unnecessary. Normal mood-related processes may be sufficient to explain the ease of depression's return.

The mood system is open to learning and history. Mood is densely interwoven in a web of thoughts, feelings, experiences, and memories. In humans, this web constitutes our autobiography. Imagine a friend returns to his childhood hometown after being away for a decade. Walking through his elementary school playground, he might well be flooded with memories of roughhousing, dodge ball, and the gruff voices of the playground monitors. If he passes by the old movie theater, other images come to mind. Without trying, he remembers the line of people around the corner on a hot summer night, the moldering smell of the lobby, and the uncomfortable seats. Wandering on his old street, he reexperiences some of the awkward pangs of his first love and the squabbles of a sibling rivalry. It all comes back. As we have all experienced, what we think, feel, and remember is strongly driven by the context at hand.

Contextual connections among our thoughts, feelings, actions, and memories do not exist to serve our nostalgic pleasure. They are evolutionarily significant assets for survival. Imagine you're a hunter-gatherer looking for food. You are in a dark forest struggling to remember the location of a patch of delicious berries. You see the felled tree that you rested on earlier in the day when you first found the berry patch. Then *BOOM!* You remember exactly where to go. As you strike out in the right direction, your memory for the exact location is sharpened by a pang of hunger. In a world where we face so many behavioral options at any given moment, close interconnections among thoughts, feelings, and memories help us shine our mental spotlight on what is most important in the here and now. Mental states like hunger are critical in prioritizing some actions, thoughts, and memories over others. You may have had an argument with a tribal elder earlier in the day, but you won't be distracted by it until you have the tasty berries in hand.[14]

Just like hunger or pain, moods are survival-relevant mental states that can bind together thoughts, feelings, and memories. An angry mood changes our mental priorities. It brings to mind every indignity we have ever suffered from a boss or spouse. We're ready to mount a strong defense at any new slight. So, too, a sad mood brings forth a history of our failures and imperfections. We see more clearly where we have gone wrong.

One technical term for these cognitive effects of mood is *mood-congruent memory*, or the increased ability to think of content that matches our current mood state. In experiments, happy subjects retrieve memories of pleasant personal experiences more readily than those of unpleasant experiences, whereas sad subjects retrieve sad memories more readily than happy ones.[15] The utility of mood congruency is that a mood will automatically cue up thoughts and memories that are most relevant to the present situation. In the case of a sad mood, we automatically bring thoughts and memories about losses to mind. Having this material at the ready fortifies us for any situation in which we're at risk of making a consequential mistake.

But mood congruency, adaptive as it may be, comes with costs. For one thing, it makes us mentally less nimble. Ordinarily we have some capacity to shift our mental spotlight, and people in sad moods commonly try to think positive thoughts or call up happy memories in order to do so. But the strong mood congruency of deep depression fixes attention on sad content, making it difficult, to say the least, for a person to shift attention away from negative thoughts and memories toward positive ones.[16] When asked about her day, a depressed client in therapy often cannot think of even a single positive event. The pleasure of finding a good parking place or getting a compliment is overshadowed by the dark cloud of low mood. In studies where we use an interview to assess a person's ability to recall personally relevant happy memories, I have seen more than one depressed person literally struck dumb. Deep depression renders people unable to recall a single meaningful happy memory from their entire lives, despite repeated prompts from a sympathetic interviewer. Perhaps not surprisingly, even when depressed people are able to call up happy memories, these memories have less impact.

Thinking about happy times does not bring the usual mood benefits.[17]

Mood congruency, over the long run, entails other significant costs. Each day on which sad material receives preferential processing is a day when the bias toward sad associations grows stronger. When days are multiplied by months and years, mood congruency spins a large and elaborate web. Concretely, research even confirms that there is a special structure in the brain, the hippocampus, where this process, known as *memory elaboration*, is centered. The hippocampus, long known to be important in memory formation, is also tasked with taking different aspects of a memory and helping to distribute them throughout the brain. My colleague, memory expert Kenneth Malmberg, explained the consequences of memory elaboration thus: "You ARE your past experiences."[18]

The hazard of being caught in this web is often hidden. As depressed people recover and return to a normal mood, their characteristic pattern of negative, pessimistic thinking recedes.[19] On the surface all is well. Yet the array of depressive associations—the interconnected thoughts, feelings, and memories—lies in wait beneath the surface, just as our forgotten childhood memories remain dormant only to be revived on a trip to our old hometown.[20] A brief spell of low mood can be enough to bring to mind the full complement of negative thoughts, feelings, and memories.[21] For someone with a strong array of associations, even a mild low mood may be enough to reopen these cognitive floodgates.

The notion that deep depression strengthens the web of sad mood is more than a gauzy idea. Controlled data support it. One representative study by Jeanne Miranda and Jacqueline

Persons examined negatively toned thinking about the self in people who had a history of depression as well as in those who had no history of depression. Negatively toned thinking, called *dysfunctional attitudes*, evidenced by statements such as, "If I ask a question, it makes me look stupid" or "If I fail at my work, then I am a failure as a person,"[22] was measured at the outset of the experiment when participants were in a benign mood state. Importantly, at this point the measurements among people with a history of depression did not differ from those of people with no history of depression. The experimenters then had each participant read a series of negative phrases, which induced a bout of low mood. The change in mood did not induce more dysfunctional attitudes in participants who had no history of depression. But for those who had depression history, the low mood triggered a spike in those sorts of dysfunctional negative thoughts.

Of course there is a great distance between endorsing a few negative statements in a laboratory and experiencing a relapse of deep depression. Do strong links between sad mood and negative cognitions actually predict relapse? Zindel Segal and his colleagues examined this issue in a study of depressed patients who had reached a stable level of remission via cognitive-behavioral therapy or antidepressants.[23] The researchers measured patients' tendencies toward negatively toned thinking about themselves before and after a low mood state was induced by listening to gloomy music or reading a series of depressing statements. Patients varied in how much they shifted toward more negative thinking after the mood induction. Some showed little change, whereas others shifted dramatically. In the next phase of the study participants were followed clinically for several years. The key finding: those who showed the largest increases in negatively toned thinking after the mood induction were the most likely to have their depression relapse during the follow-up period.

Breaking Free

The mood system has a bias to return to deep depression even with little provocation. Fortunately relapse is not inevitable, and it can be countered. Antidepressant medication is currently the dominant strategy for buffering a person's risk of relapse. Using antidepressants as the first line of defense is consistent with defect models, such as the biological model of kindled depression. In line with the idea that the drugs address a permanent vulnerability, psychiatrists often recommend a *lifetime* of antidepressant maintenance treatment for people who have previously experienced three or more episodes.[24]

But antidepressants aren't the only proven means for slowing down or preventing a relapse. A relatively brief course of a psychologically based treatment—the best examples being cognitive therapy and mindfulness-based cognitive therapy—demonstrates comparable protection.[25] The success of psychologically based treatments has implications for how we think about vulnerability to depression.

The tendency for depression to repeat reflects the normal default settings of a plastic mood system that is open to experience. The unfortunate consequence of plasticity is that a long duration of deep depression can reprogram the entire system so that it favors a return to low mood states. The good news is that there is a flip side to plasticity. As we see with these psychologically based treatments, the mood system can be deprogrammed. Mindfulness-based cognitive therapy is particularly intriguing, because it purports to work by de-coupling sad mood from the tendency to engage in nega-tively toned thoughts about the self. Synthesizing techniques drawn from Eastern religion and cognitive therapy, mind-fulness-based cognitive therapy uses meditative practices to help a person tolerate and accept episodes of low mood while avoiding an endless loop of negatively toned cognition. We have much to learn about exactly why mindfulness works (when it does), but the supportive data suggest that specific mental techniques can arrest the downward spiral of mood. The success of brief, psychologically based treatments is en-couraging not only because the treatment works, but because it speaks against the existence of a permanent, brain-based vulnerability to depression.

Thus far I have tried to explain why it seems that we are losing the fight against depression. More people are be-coming depressed. Despite available treatments, depression is difficult to contain, running a recurrent or often chronic course. Naturally I have focused on the liabilities that explain these bad outcomes. Some liabilities emerge when an ancient mood system finds itself in a novel operating environment, others are fed by the sometimes peculiar ways that our spe-cies regulates mood, and still others emerge from emphases in our contemporary culture. These liabilities have conspired to provide a hospitable environment for low mood.

But our fight against depression isn't a lost cause; far from it. And I am all for hope—when it's realistic hope, the right amount of hope—because several of the things that make people vulnerable to depression can be changed. It is possible to alter the trajectory of mood by modifying daily routines, resetting goals, and changing how we interpret or react to moods. It is less feasible to change culture or a person's temperament, but we can become more aware of how these factors impact mood, itself a clearly positive step.

Furthermore, though overall trends in the rates of depression onset and recurrence are quite bleak, it is important to stress that many people buck the odds. Not everyone who becomes depressed stays in limbo, and not everyone has a recurrence. In fact, by some recent estimates half the people who have a first episode of depression will never have another.[26] Although we lack a full accounting of these people's experiences, we know many will recover and stay well without engaging in any formal therapy. As we turn to this brighter side of breaking out of limbo and staying well, my own case suggests that a focus on liabilities may not tell us everything we need to know.

In the course of writing this book, I decided the time was right to tell my sixteen-year-old daughter Sophie about my history: that I had suffered from depression and that it had been bad enough to nearly swallow me whole. Despite my anxieties about broaching the topic, the conversation was surprisingly natural and matter of fact. Toward the end Sophie asked me, "Are you worried it's going to come back?"

I hesitated before I answered. By the numbers, I should worry. My liabilities stack up. My depression came on in young adulthood. Generally speaking, the earlier in life the first episode occurs, the worse the expected future course is. My episode was long, more than four years. Again, generally speaking, longer episodes portend a worse course. Four years of deep depression almost surely changed my mood system, fusing contexts, thoughts, feelings, and behaviors into a mental mode of a depressed Jon. Perhaps he's still there, waiting to be resurrected. Further, my episode was severe and impairing, also foreboding for the future.

Another liability was that my depression didn't respond to a host of treatments. It was resistant to half a dozen antidepressants, as well as to booster drugs that were brought in to augment the antidepressants, and to a month-long psychiatric hospitalization at Johns Hopkins University's respected affective disorders unit.[27] Finally, I have a positive family history for depression. There can be little doubt I carry around a genetic liability for it. Growing up, I heard stories about depression among many relatives on my father's side, even going back to the Old Country, Russia; my mother's side, though not as rich with depression, has relatives affected with anxiety as well as schizophrenia.[28] I have a moderately neurotic temperament that probably owes itself to my family lineage.

Given these factors, I should have relapsed already. Yet in thirteen years there has been no recurrence of deep depression for me. And the more time passes without a relapse, the more likely it is that I will be among those who have a single lifetime experience of depression. (If all goes well, I will be in Scott Monroe and Kate Harkness's memorable acronym, a SLED: single lifetime episode of depression.)[29] I told Sophie, "At this point, I'm not too worried. I think I'm going to be one of the lucky ones."

What I told Sophie is true. I believe I am out of the woods—and I do feel lucky. As a dad, I'm happy to be able to tell her all this, but as a scientist I'm not completely satisfied with my statement. My training makes me uneasy with a happy mystery, or the idea of dumb luck. It would be better to have a real explanation for my good fortune, ideally one that wasn't unique to me, but could apply to other people who have achieved a full recovery from deep depression despite their liabilities. In the final chapter I consider what is known about those who achieve these good long-term outcomes.

CHAPTER 10

The Glory of Recovery

DEEP DEPRESSION IS CRUEL. THE AFTERMATH OF AN EPISODE IS A dull hangover of residual symptoms. Episodes of deep depression are prone to recur, sometimes over decades. Yet for a subgroup of sufferers, depression does not cloud the future. For this group, mental health is regained. Depression does not and may never return in any serious way.[1] For some the recovery is truly glorious: life after depression is *better* than it was before depression. Who are these people? Why do they experience these uniquely good outcomes? Though these are important questions, not much research has been devoted to them.

Fortunately mood science gives us some tools for thinking about these issues. Doing so will require us to change gears. To this point we have naturally focused on the many internal and external factors that pull a person's mood down. Now we consider the flip side, the ways in which the mood system, as the great integrator, is affected by processes that build up mood.

As we consider the possibility of a glorious recovery, it's natural to wonder if reading this book can help you recover from your depression. Before I speak to this, I have a few caveats. First, this book is no substitute for advice from a mental health professional. Another caveat follows from my main premise, which is that we need to go back to the drawing board. We are losing the fight against depression in part because our fundamental description of it—as reflecting defects—is wrong. The first step to finding more effective solutions is getting that fundamental description right, and my book is one effort toward that end. Finally, I am skeptical of any easy, one-size-fits-all solution for depression, and you should be, too. The genre of self-help for depression is littered with well-intentioned books that overpromise solutions and raise false hopes. It would be nice to defeat your depression in ten easy steps, but rarely is it so easy. Books that overpromise solutions produce frustrated, disappointed, and demoralized readers and damage the credibility of experts.[2] I haven't written a self-help book, or at least not in the usual sense.

These caveats out of the way, let me affirm that the mood science approach *does* have relevance for individuals who are struggling with depression. It *can* offer insights about the recovery process. And in some ways, it's actually *more optimistic* than competing approaches. As we will see, not only is a glorious recovery possible, but mood science can help explain *why* it happens.

One advantage of the mood science approach is its breadth. Because the mood system is the great integrator, it is open to multiple inputs. I have shown the multiple ways that people, often unwittingly, pull down their own moods.

The opposite holds as well. It's an optimistic fact that there are multiple ways by which mood can be rebuilt. Help may come from many quarters. We can alter mood in an upward direction by changing how we think, events around us, our relationships, what is happening in our bodies (by exercising or sleeping better), and our brains through medications or diet. In this sense, the mood science approach invites the sufferer to browse all the depression books in the store.

Suzie Henderson's long journey out of depression exemplifies the benefits of—and need for—a diversity of countermeasures. She struggled with depression for six years in her teens and twenties. She tried antidepressants and several therapists, without much luck, before piecing together a solution to her mood problems. She wrote in a detailed account of her recovery:

> *A doctor recommended I take fish oil, so I started taking it, and didn't really notice changes until a couple months of consistent use, and then I perceived a more even keel attitude shift within myself. I started meditating, and doing yoga. I began slowly, maybe a class once a week. I started educating myself about nutrition and eating balanced meals. Here is a key ingredient to my recovery, I am constantly observing my thoughts and using positive self talk. I use a mantra when my mind chatter is especially negative, and it's really simple. I just remind myself that I am strong, healthy, and beautiful. I say it many times throughout the day, not necessary to keep count. . . . I have done it for years now. Basically crowd out negative habits with positive ones. It is not a quick fix, but if you have the patience. . . . [I]t has really helped me. Other helpful methods, getting to know*

myself and how I work, and working to achieve balance in my life. Enough sleep, de-stressing, eating well, reminding myself of the gifts I have, and everything I am grateful for. You know it's easy to focus on the negative, but even keeping a journal where you just write about things you noticed during the day that stirred a sense of gratitude inside you is helpful. Maybe it comes down to retraining the mind and body. Support is important, friends, therapists. Something that also helps me is working towards a goal and actualizing my potential.

As illustrated by Suzie's story, many levers can move mood. Deliberate experiments with pulling on the various levers can pay off. In addition to trying formal therapies and medications, Suzie took the initiative to change her diet, sleep, cognitions, and relationships with others. Conventional approaches to depression, by contrast—whether mainstream psychiatry or cognitive-behavioral therapy—tend to present a small subset of the available levers as the sole solutions. Holding no approach sacred, mood science is guided by the openness of the mood system. The sufferer is empowered to improvise, to make his or her own recovery.

Suzie's experience also raises the question of what recovery should look like. Disease/defect models define recovery by what is not there. The end of depression is an absence of illness. The idea behind standard treatment approaches is that by correcting defects in the mind, brain, or interpersonal functioning, the depression symptoms will abate. It's a return to the status quo ante. The slate is wiped clean. This conception of recovery may seem sensible enough, but defining recovery by an absence is both vague and a bit hollow. An alternative conception might define recovery more positively and encompass ideas of wellness or thriving.[3] As seen in Susie's case, thriving or wellness could involve several factors, such as personal growth, subjective well-being, or increased purpose or meaning in life.[4]

Wellness is an attractive goal. Most people, if they could choose what their recovery would look like, would surely prefer wellness to being "not ill." Is correcting defects sufficient to bring about wellness? Probably not, but this question is rarely asked in the treatment of depression. The vast enterprise of depression treatment research, inspired by defect models, has virtually nothing to say about wellness or thriving.[5] The bottom line is that we know shockingly little about those who defy the usual odds and thrive after depression. One even wonders if this silence reflects an unspoken view that depressed people should not bother to aim so high.

The mood science approach takes the aspiration of thriving seriously. This is not merely on aesthetic grounds, or because it's what depression sufferers might want in a perfect world. It's because wellness and thriving may prove to be more robust and clinically meaningful endpoints than the simple absence of symptoms. In one large survey, researchers took various measurements of positive well-being as well as a standard measure of depressive symptoms. The key finding was that an absence of well-being was a strong predictor of future depression and was more predictive than the symptom measure. In fact, the absence of positive well-being was associated with a sevenfold increase in significant depression ten years later.[6] This is only one study, of course, but its results point to the possibility that these various elements of well-being may be a better way to index recovery, and a better means of predicting future functioning, than standard-issue depression symptom measures.

Finally, the mood science approach takes the goal of wellness or thriving seriously because it is both a fascinating and a legitimate research question to investigate how people thrive after depression. Yes, we start at a disadvantage because the topic has been ignored. But by blending together what we know about key ingredients of human flourishing[7] with individual stories of depression sufferers who achieved wellness, we can extract some key themes and underlying lessons.

Postdepression Personal Growth: Sylvie's Story

In the worst phase of her depression, Sylvie was overwhelmed by responsibilities at work and home and was overpowered by brutal symptoms. She felt trapped. After planning it for days, one Friday during lunchtime she went to a pretty lake and took forty-five sleeping pills. When she woke up in a hospital bed after the failed suicide attempt, her first thought was, "Oh my god, I fucked up," and then, "How come I'm not dead?" Sylvie had no particular reason to think she would ever recover, let alone thrive.

Sylvie's recovery proceeded in stages, and like Suzie Henderson's, involved multiple changes in her life. As discussed previously, in the months after her release from the hospital Sylvie gradually shed her severe physical symptoms. Her limbs had been heavy. She couldn't swallow, and she had lost thirty pounds without trying. It had been difficult to move, a struggle to even open the mailbox. She could not sleep. Sylvie credits antidepressant drugs for helping her cope with her physical symptoms. As she put it, "They gave me my body back." Being able to sleep was "the most magnificent thing in the universe." Once set back in motion, Sylvie was able to restart her life. She started to recognize herself. She returned part time, then full time to her job as a counselor. Her colleagues knew she had been depressed but knew nothing of the suicide attempt. With work anchoring her, she started to process the guilt and shame of her failed attempt, even though she could not yet talk about it to others.

But for Sylvie to really turn the corner and come back feeling "better than ever," she needed to enter a new phase of recovery. She used her shattering depression as a lens to reevaluate everything in her life, figuring out what was most important to her, and making changes to her life in light of that reevaluation. In other words, she started on a path of postdepression personal growth. This idea might seem odd, and surely it does not fit with the disease model, in which depression produces nothing of value. Yet it is increasingly recognized in other areas of research that people often achieve growth amid pain. For example, in the area of trauma there is an increasing amount of systematic research showing that a sizeable subgroup of the traumatized population achieves post-traumatic growth. After surviving a horrible accident, people routinely report a greater sense of personal strength and/or appreciation of life than before the event.[8] This research shows that some traumatized people actually find benefits amid their unfortunate experiences and find ways to make themselves better for their struggles.[9] Examples of post-traumatic growth might include renewed appreciation for life, seeing new possibilities for oneself, feeling more personal strength, improving relationships, and feeling spiritually more satisfied. So, too, the hardship of depression should present similar opportunities for personal growth.

For Sylvie, postdepression growth involved dramatic new evaluations of herself and her life. She eventually came to see her depression not as a wound or vulnerability but as a yardstick, by which she came both to know and understand her true self. Sylvie complained that before her depression she often lived in fear of offending others or of having something bad happen to her or others. Since the depression she is more comfortable with who she is and lives with less fear: "I will honor myself and not betray myself or feel shame about who I am anymore." She continued:

> My depression has allowed me to declare who I am and screw anybody who doesn't like it or approve of me. No one's opinion of me matters. I no longer seek approval. AND I will go the extra mile to reassure others, to do no harm, to own my own behavior and apologize first and deeply because I recognize how frail every one truly is and I must not push anyone else near their tipping point in my life because it is way too dangerous and difficult to come back. I am much more comfortable being wrong and backing off of my point when I see someone getting upset. I will just shut down the discussion or conflict etc. It is never worth it. Forgiveness comes easily and so does acceptance—no judgment of others. I am so much more loving to others, and I watch out for myself fiercely.

In the words of Frederic Flach, a pioneer in recognizing the potentially transformative power of depression, the process of overcoming depression has given Sylvie "a secret strength."[10] After the turmoil of her depression, she perceives "a new inner peace that was not available before." The usual hassles and problems of the world seem smaller. This peace, almost paradoxically, has also made her more willing to assume risk. After depression, she said, "I am willing to be vulnerable and embrace the natural flow of life rather than trying to direct it to my own course and yet it has given me new courage because there is no consequence that could come as close as wanting to die. . . . [T]he very worst thing that can happen in a life is wanting to end it. So I live more bravely than ever with more respect for others and myself."

Not only did depression lead Sylvie to new understandings; it also led her to change her behavior. Depression helped her grow in her work; she is a better social worker for it. She uses her experience of depression to achieve greater empathy and efficacy with her older patients, who face serious problems such as a diagnosis of Alzheimer's disease. She spots depression in others easily and is in a strong position to deliver realistic hope. When others sound desperate and suicidal, she says she gives them the "gift of my depression," which is her knowledge that "you don't have to try to kill yourself to become a crusader for your life."

Another paradox is that her experience of terrible pain during depression has made it easier for Sylvie to appreciate, and savor, normal pleasures like spending time with her daughter, Madeline. As Sylvie put it, "Every day is a gift to be celebrated. The time to do fun and exciting things is NOW." Someone who previously could not enjoy anything now finds pleasure almost everywhere, and her enthusiasm is infectious.

Even though there is virtually no research on postdepression growth, I am confident that Sylvie's experience is mirrored in other recoveries. After all, a central function of depressed mood is to slow behavior and prompt reevaluation when efforts toward key evolutionary goals are not paying off. Successful reevaluations should routinely lead to post-depression growth. Thoughts in depression need not revolve in an endless loop of rumination, in which the same negative ideas are turned over and over. For Sylvie and others, deep depression can instead be a kind of creative destruction—a forced questioning of basic assumptions—leading to true re-evaluation, new meanings, new beliefs, new goals, new behaviors, and even a new life narrative. As Sylvie sums up her recovery, "I'm like me, but better."

Upward Spirals of Well-Being

A second characteristic of sustained and strong recoveries is the experience of elevated well-being. I know this sounds circular. Depression ends when the person feels better. But it is not as circular as it sounds. Remember our study showing that the absence of well-being predicted future depression ten years later? Well, the obverse of that finding is that the presence of well-being predicted who *wouldn't* have future depression. This suggests that positive emotions are not simply an outcome, but an active process that can sculpt the future.

One of our key premises is that moods, positive as well as negative, have functions. Positive moods are not only a sign or readout that we are on the right track and moving toward evolutionarily favored goals; they also feed forward into our future behavior, shaping the choices we make and how vigorously we pursue these goals. In other words, we need to understand how the experience of well-being might help people *do* things that keep them well.

Fortunately a rich vein of research consistently links the experience of positive emotions to a variety of positive life outcomes. In an influential series of review articles, Sonia Lyubomirsky has cataloged the ways that happy people appear to benefit from their positive states of mind in many different life domains.[11] Over the long run, happier people garner more friends, enjoy stronger social support and richer social interactions,[12] and have increased productivity at work and earn higher incomes.[13] Of course the critical question is, why?

Barbara Fredrickson, a psychologist at the University of North Carolina, has done extensive work to unlock this mystery, to explain why the experience of positive moods might be linked to other enduring benefits. Her broaden-and-build model of positive affect focuses on the ways that it functions to broaden attention and build resources.[14] The functions of positive mood are in essence the opposite of the functions of low mood and negative emotions. If negative emotions such as anxiety narrow the focus on a threat (a key problem in adaptation), positive emotional states do the opposite: they broaden focus on new opportunities and help to build a variety of personal resources—psychological, cognitive, social, and physical. And it's ultimately these resources that contribute to well-being over the long haul. Consistent with this theory, experimental evidence has shown that positive emotions broaden what Fredrickson calls *thought-action repertoires*. For example, when positive emotions are induced in a laboratory setting, participants evidence wider visual search patterns, display novel and more creative thoughts and actions, and are more flexible in their goals and mindsets. Fredrickson and her colleagues explain that over time these states have real-world consequences: "[I]dle curiosity can become expert knowledge," "affection and shared amusement can become a lifelong supportive relationship. Positive emotions forecast valued outcomes like health, wealth, and longevity because they help build the resources to get there."[15]

This theory has clear application to the aftermath of depression. Frequent experience of positive emotions helps explain why a recovery from depression can become self-sustaining. Positive emotions push people to engage in novel and exploratory behaviors such as making a new friend, looking for a job in a new industry, or developing a new hobby, all of which may lead to new payoffs. It is precisely this ability to strike out in a new direction, to capitalize on an opportunity, that is so badly needed in the aftermath of deep depression. The depressed person has just been through a long winter of inflexible thinking and behavior. Positive emotions allow for a critical thaw.

Much like the old story of needing to have credit to get credit, demonstrations of the dynamic power of positive moods put the sufferer in a frustrating position. *If all I need to do to stay well is to feel well, let's get me there!* Although this impatience is understandable, it's important to reiterate that experiencing well-being during recovery, or at other times, is not simply a matter of wanting to feel good. I have discussed how the goal of feeling happy is unusual; it's not like other goals such as learning to bake a pizza, for which the desire to achieve is half the battle and steady application is the rest. Pressing harder on unfulfilled and unrealistic goals for happiness can paradoxically deepen depression. This is why "how to be happy" books don't work like magic for depressed people. Yes, it's possible to increase well-being, but the indirect approach is often best. To paraphrase John Lennon, increased well-being is what happens to you when you're busy making other plans.[16] These other plans include things like working toward personal growth, tending toward others, as well as finding purpose in life, the topic of the next section.

Recovery from Depression and Purpose in Life

Psychologists Patrick McKnight and Todd Kashdan define *purpose in life* as a system "that sustains and builds well being." Purpose is a sort of glue that binds together actions over time. They write, "Purpose is a central, self-organizing life aim that organizes and stimulates goals, manages behaviors, and provides a sense of meaning. Purpose directs life goals and daily decisions by guiding the use of finite personal resources. Both higher-order and lower-order goals come from a purpose. We expect that people who have a purpose in life would move seamlessly from goal to goal or manage multiple goals simultaneously."[17]

The concept of purpose may sound terribly abstract and out of place. Yet one clue that purpose may be crucial for building a strong recovery is that depressed people typically face a crisis of purpose. I was not atypical in this respect. At the time I became depressed, I was dead set on a career in history and committed to the monastic life of the history doctoral student. Careers, as they are connected to economic and social resources and ultimately social status, are an important evolutionarily favored purpose, one that the mood system will carefully track. I fully defined myself by my potential career.

My depression probably originated in part out of struggles to develop a dissertation topic and from the ominous signs of a poor academic job market. Once I became significantly depressed, my ability to work and think about history deteriorated. I could no longer do the only thing that I thought I could do well. My original life plan collapsed.

It was not a simple matter to re-create purpose or even to divine the meaning of my depression—both took years. Yet it's clear that my doing well after depression has been connected to a restoration of purpose, or more accurately, purposes. To the extent that my depression offered a warning, I think it was about the hazard of putting all one's eggs in a single basket. I stay well in part because I have diversified my portfolio, evolutionarily speaking. With research psychology, yes, I re-created a career. But I also married, had a daughter, and even developed full-blown hobbies like marathon running. And for the past few years I have been writing a book that aims to help others understand depression. Each of these enterprises has given me purpose. Because each purpose can be connected to a key evolutionary theme, such as attachment, procreation, health, and affiliation, no doubt the state of each enterprise is tracked by my mood system.

The specific enterprises that will create purpose in life will differ from person to person and emerge from his or her history and needs. Your mileage will surely vary. There's no ready-made formula for discovering and rebuilding life purpose (or purposes) after depression.[18] It can and should emerge over time from solo reflection, as well as from conversations with spouses, friends, and therapists. This diverse process is worth pursuing: I expect what is common among people is that however purpose is created, it can hold depression at bay. I still have my depression-prone temperament and a set of genes that pull for low mood, and life is as stressful as it ever was. But purpose is like a talisman, a charm that can ward off serious depression. This again is a reminder that we may be better off if we think about recovery, not simply as the absence of depressive symptoms, but as a set of active qualities or practices that prevent low mood from taking root, despite the presence of liabilities elsewhere.

ONE OF THE THINGS that worries me most about the sway of conventional disease/defect models is that they systematically discourage people from trying to make sense of their depression. We have been taught to listen to Prozac, but not to depression. In fact, the stance of conventional defect models is that there is little value in listening. Depression's symptoms are only noise—the static of neurotransmitters or faulty thinking—to be silenced by drugs or cognitive therapy.

Ovid wrote, "Welcome this pain, for you will learn from it." That is surely too glib. Our perspective here is that, although depression's pain is never entirely welcome, moods offer meaningful information about our status and prospects in the world. Without trivializing how difficult it will be to "listen to depression" to extract evolution's warnings, to find the signal amid the pain, this listening, particularly in its aftermath, can be a vehicle to foster rebirth and transformative life change. Certainly it will be difficult to learn from depression if we don't listen at all.

Toward an Adult
National Conversation About Depression

In presenting the prospect of glorious recovery, this chapter may sound like happy talk. It is not. A sober-eyed view of depression includes the reality of positive outcomes, the recognition that these outcomes don't occur at random, the desire to discover the reasons for these good outcomes, and the goal of applying this information so others can build a strong, enduring recovery. If we take the perspective of evolution seriously, we will accept that we are not designed to be happy. Survival and reproduction are the ends of evolution, and low mood often serves those ends as well as happiness does. Low mood is so strongly selected for that any treatment that could rapidly eliminate it would be both highly unlikely and highly dangerous (like the major drugs of abuse, which hijack our normal reward systems for other ends). Low mood is both inescapable and sometimes useful. Thus, an evolutionary perspective asks us to be patient, to learn to tolerate some degree of low mood, and to listen to what it is that low mood can tell us. There are no guarantees, but becoming happier, reducing the sway of low mood, and ultimately thriving are all realistic goals if we manage our moods wisely and move our lives in directions favored by evolution. There is no reason that outcomes of depression must continue to be as bleak as they now are. Yes, we are losing the fight against depression—but this does not mean that all is lost.

Indeed, my great hope in writing this book is to restart our national dialogue about depression. Our national conversation for the last twenty years has been on hold, largely reduced to a narrow dialogue about the promise and peril of antidepressants: "To Prozac, or not to Prozac?"[19] Peter Kramer's *Listening to Prozac* raised expectations that antidepressants would soon make depression obsolete. As this proved not to be the case, there was the predictable backlash. In Robert Whitaker's *Anatomy of an Epidemic*, antidepressants (and other psychotropic medications) are not only ineffective but the villain, responsible for worsening the epidemic of mental illness. Antidepressants remain an eight-hundred-pound gorilla of mental health, so we can predict that soon there will be a backlash to the backlash, with new arguments advanced to restore the luster of drugs. And so it goes. But the truth is that antidepressant medications are only one way to move the mood system, and it turns out that they are not an especially potent way of doing this. In Irving Kirsch's exhaustive analysis of clinical trials, inert placebo pills are about 82 percent as effective as antidepressants.[20] Of course Kirsch, because he wrote on this topic, naturally attracted a backlash. And that's exactly the point. If we do nothing different, we can conclude with supreme confidence that all the heated talk about drugs will continue to monopolize the stage and preclude a real conversation about mood and mood disorders.

I make this plea for an honest, balanced, and adult national conversation about depression. We need this not only for the adults. We also desperately need it for my daughter, Sophie, and for the rest of her generation, the teens who will soon be young adults. Our youth will face depression in high schools and on college campuses in epidemic proportions that will overwhelm them, their parents, and all counseling resources. As it stands, the young have only limited means to cope; they currently lack any strong sense about how low mood or depression work or what to do about either, aside from possibly taking medications. Ironically, our culture is obsessed with mood, with feeling good, yet, in part because of our impoverished fixation with depression as a disease, it continues to be mood illiterate, ignorant of its many real sources. For many, dental hygiene is taken more seriously than mood hygiene.

Finally, accepting depression as a legacy of our animal nature is an important step toward making our national conversation about depression not only realistic, but humane. It is high time for our society to revise its stance toward the millions who have battled depression. The conventional approach tends to view the legions of the formerly depressed as a "broken" people, an ever-afflicted group that will likely need repeated assistance over the life course because of their theorized defects. This stance is inaccurate and belittling, and its unwitting effect is to continue the corrosive, age-old stigma attached to depression and depressed people, albeit in a slightly different form.

One obstacle to a more affirmative national conversation is that depression has lacked a unifying public symbol that could bring it out of the dark, as Livestrong© bracelets did for cancer or the rainbow flag did for LGBT. When most people think of depression, their first associations are to unfortunate images, such as a dark cloud, the color black, or a noose. One reason why depression stigma lives is that depression has a serious bumper sticker problem.

But this is essentially an issue of failed marketing and messaging. It should be possible to develop a unifying symbol, and if it is presented in a compelling way, many might rally (see Figure 10.1). Conservatively, thirteen million US

FIGURE 10.1. Come Out of the Dark
Wristband, Proposed by the Author as a Rallying
Point to Fight Depression Stigma.

Photo credit: Sophie Rottenberg

adults are currently in an episode of depression; more than twice that number have had depression in the past. When we add in caregivers, millions more are indirectly affected by the quality and the quantity of our national dialogue about depression. Have no illusions. Even with a strong public education campaign, stereotypes that have been decades in the making will resist rapid change. Still, with so much to gain, it is high time we try.

But in my view, finding more humane ways to discuss the predicament of depressed people is not just good marketing, it's also good science. The mood science perspective tells us that depression, deep or shallow, is a natural product of the mood system. However a person gets there, facing deep depression is a supremely difficult trial. Rather than assuming weakness or defectiveness, we should acknowledge that getting through depression requires considerable strength. Rather than assuming permanent debility, we should recognize that some depressions are followed by thriving. Writing these words fourteen years after my episode, I recognize that I am not broken. Getting beyond the disease model will require us to honor the strengths of formerly depressed people, to see their potential for rebirth after depression and the ways that, once reborn, they can help others build enduring recoveries from depression.

It is possible.

Acknowledgments

THIS BOOK HAS BEEN A LONG TIME IN COMING, AND IT WOULD have been much longer were it not for all the help I had along the way.

First, I would like to thank Ian Gotlib and James Gross for their generous professional support, as well as the Psychology Department at Stanford University for their willingness to take a chance on a failed historian. Early on, Sheri Johnson and Ann Kring helped me see the excitement and ferment in the field of emotion and psychopathology, and Randolph Nesse helped make room for books like these, by fighting for evolutionary approaches to mental disorders when it was unfashionable to do so.

Several years ago I had the good fortune to receive an e-mail from Max Brockman suggesting that I might have a book in me. My agent, Lisa Adams, helped me hone my proposal, and at every step I have benefited from her patience, editorial acumen, and unfailing good sense. I was fortunate that Lisa helped match me with the strong team at Basic Books, including TJ Kelleher and Tisse Takagi, who gave me trust and freedom to roam and helped steer whenever I got off course.

In preparing the manuscript I benefited greatly from the work of talented research assistants Penny Carlton and Deborah Duey. Ena Begovic provided truly herculean labors over eighteen months, tolerating my disorganization, tracking down sources, and developing many of the figures that appear in this book.

The Mood and Emotion Laboratory has been a great forum to develop my ideas. Beth Morris, April Clift, Vanessa Panaite, Lauren Bylsma, Megan Howard, John Grace, and Bethanne Bower all provided helpful comments on chapter drafts. I have appreciated the consistent support of colleagues in the Psychology Department at the University of South Florida for the past ten years, including the "Psychos" group.

Much of the manuscript was written in the Netherlands during a sabbatical year at Tilburg University. I thank Ad Vingerhoets for his warm invitation to come to Holland and for the opportunity to collaborate with him. In Maastricht, Frenk Peeters provided friendship, running, and great conversations about depression, and Marjolein Selis periodically fed and cared for us and our dog and made us laugh. My American colleagues Renee Thompson, Douglas Mennin, and James Gross read and commented wisely on portions of the manuscript.

I am grateful to Nancy Reiley, Robert Rottenberg, and Rana Rottenberg for their unfailing moral and emotional support; to Sophie for being Sophie; and to Ollie and Cy for furry companionship. Finally, this book is dedicated to my wife, Laura Reiley. She knows why.

Notes

Chapter One: Why We Need a New Approach to Depression

1. Kessler, R. C., Berglund, P., Demler, O., Jin, R., Koretz, D., Merikangas, K. R., . . . Wang, P. S. (2003). The epidemiology of major depressive disorder: Results from the National Comorbidity Survey Replication (NCS-R). *Journal of the American Medical Association, 289,* 3095–3105.

2. Bromet, E., Andrade, L. H., Hwang, I., Sampson, N.A., Alonso, J., de Girolamo, G. . . . Kessler, R. C. (2011). Cross-national epidemiology of DSM-IV major depressive episode. *BMC Medicine, 9,* 1–16.

3. Lopez, A. D., & Murray, C. C. J. L. (1998). The global burden of disease, 1990–2020. *Nature Medicine, 4,* 1241–1243; World Health Organization. (2008). *The global burden of disease: 2004 update.* Geneva: World Health Organization.

4. For the 30 percent increase, see Centers for Disease Control and Prevention, *Morbidity and Mortality Weekly Report (MMWR),* http://www.cdc.gov/mmwr/preview/mmwrhtml/mm6217a1.htm?s_cid=mm6217a1_w.

5. The widely cited national comorbidity study puts lifetime prevalence of major depression at 17.1 percent. Blazer, D. G., Kessler, R. C., McGonagle, K. A., & Swartz, M. S. (1994). The prevalence and distribution of major depression in a national community sample: The National Comorbidity Survey. *American Journal of Psychiatry, 151,* 979–986. The national comorbidity survey replication puts the prevalence at 16.2 percent.

6. One article reports more than a doubling of depression in a large sample over ten years! Compton, W. M., Conway, K. P., Stinson, F. S., & Grant, B. F. (2006). Changes in the prevalence of major depression and comorbid substance use disorders in the United States between 1991–1992 and 2001–2002. *American Journal of Psychiatry, 163,* 2141–2147. There is extensive information on the increase in depression among younger birth cohorts in Burke, K. C., Burke, J. D., Jr., Rae, D. S., & Regier, D. A. (1991). Comparing age at onset of major depression and other psychiatric disorders by birth cohorts in five US community populations. *Archives of General Psychiatry,*

48, 789–795; evidence that this is an international phenomenon in Cross-National Collaborative Group, Weissman, M. M., Wickramaratne, P., Greenwald, S., Hsu, H., Ouellette, R., . . . Hallmayer, J. (1992). The changing rate of major depression: Cross-national comparisons. *JAMA, 268*, 3098–3105. Conversely, older cohorts have less depression than would be expected otherwise; one cross-sectional study found that individuals born in the middle third of the twentieth century were ten times more likely to suffer major depression than individuals born in the first third of the century. Klerman, G. L., Lavori, P. W., Rice, J., Reich, T., Endicott, J., Andreasen, N. C., . . . Hirschfield, R. M. A. (1985). Birth-cohort trends in rates of major depressive disorder among relatives of patients with affective disorder. *Archives of General Psychiatry, 42*, 689–693.

7. Pescosolido, B. A., Martin, J. K., Long, J. S., Medina, T. R., Phelan, J. C., & Link, B. G. (2010). "A disease like any other"? A decade of change in public reactions to schizophrenia, depression, and alcohol dependence. *American Journal of Psychiatry, 167*, 1321–1330, appi.ajp.2010.09121743.

8. Matt, interview with the author, November/December 2010.

9. For a book-length treatment, see Finnigan, W. J. (2008). *The demon of depression.* n.p.: Xulon Press.

10. Faced with bewildering symptoms, including concentration problems, depressed people typically have little confidence in their own interpretations of their mood, which partly explains why sufferers appeal to experts.

11. Dr. David Goldbloom on *The Agenda*, aired May 17, 2007.

12. Dr. Brenda Smith on *The Agenda*, aired May 17, 2007.

13. Reported in Olfson, M., & Marcus, S. C. (2009). National patterns in antidepressant medication treatment. *Archives of General Psychiatry, 66*, 848–856; for evidence of indiscriminate overmedication, see Mojtabai, R. (2013). Clinician-identified depression in community settings: Concordance with structured-interview diagnoses. *Psychotherapy and Psychosomatics, 82*, 161–169.

14. Trivedi, M. H., Rush, A. J., Wisniewski, S. R., Nierenberg, A. A., Warden, D., Ritz, L., . . . STAR*D Study Team. (2006). Evaluation of outcomes with citalopram for depression using measurement-based care in STAR*D: Implications for clinical practice. *American Journal of Psychiatry, 163*, 28–40.

15. Curry, J., Silva, S., Rohde, P., Ginsburg, G., Kratochvil, C., Simons, A., . . . March, J. (2011). Recovery and recurrence following treatment for adolescent major depression. *Archives of General Psychiatry, 68*, 263–270.

16. Beck, A. T. (1979). *Cognitive therapy of depression.* New York: Guilford Press.

17. Dismantling the active ingredients of a therapy is a challenge in the field. Jacobson, N. S., Dobson, K. S., Truax, P. A., Addis, M. E., Koerner, K., Gollan, J. K., . . . Prince, S. E. (1996). A component analysis of cognitive-behavioral treatment for depression. *Journal of Consulting and Clinical Psychology, 64*, 295–304.

18. Carey, B. (2010, January 5). Popular drugs may aid only severe depression, analysis says. *New York Times.* Retrieved from http://www.ny times.com/2010/01/06/health/views/06depress.html?_r=1. Fournier, J. C., DeRubeis, R. J., Hollon, S. D., Dimidjian, S., Amsterdam, J. D., Shelton, R. C., & Fawcett, J. (2010). Antidepressant drug effects and depression severity: A patient-level meta-analysis, *JAMA, 303,* 47–53.

19. Stacey Murrette, interview with the author, November 4, 2010.

20. American Psychiatric Association. (2013). *Diagnostic and statistical manual of mental health disorders: DSM-5* (5th ed.). Washington, DC: American Psychiatric Publishing.

21. Books that marked the coming together of affective and clinical science include Rottenberg, J., & Johnson, S. L. (Eds.) (2007). *Emotion and psychopathology: Bridging affective and clinical science.* Washington, DC: APA Books; and Kring, A. M., & Sloan, D. S. (2009). *Emotion regulation and psychopathology.* New York: Guilford Press.

22. This discussion synthesizes several ideas about the functions of affect. Cosmides, L., & Tooby, J. (2000). Evolutionary psychology and the emotions. In M. Lewis & J. M. Haviland-Jones (Eds.), *Handbook of emotions* (2nd ed.) (pp. 91–115). New York: Guilford.; Russell, J. A. (2003). Core affect and the psychological construction of emotion. *Psychological Review, 110,* 145–172; Morris, W. M. (2000). Some thoughts about mood and its regulation. *Psychological Inquiry, 11,* 200–202.

23. We cannot be certain that the goat feels good when it eats, though it seems extraordinarily likely to be so. Balcombe, J. (2006). *Pleasurable kingdom: Animals and the nature of feeling good.* New York: Macmillan.

24. Examples are drawn from Rottenberg, J. (2005). Mood and emotion in major depression. *Current Directions in Psychological Science, 14,* 167–170.

25. Mogg, K., & Bradley, B. P. (1999). Some methodological issues in assessing attentional biases for threatening faces in anxiety: A replication study using a modified version of the probe detection task. *Behaviour Research and Therapy, 37,* 595–604.

26. Fredrickson, B. L., & Branigan, C. (2005). Positive emotions broaden the scope of attention and thought-action repertoires. *Cognition and Emotion, 19,* 313–332.

27. Kahn, B. E., & Isen, A. M. (1993). The influence of positive affect on variety seeking among safe, enjoyable products. *Journal of Consumer Research, 20,* 257–270.

28. Cosmides, L., & Tooby, J. (2000). Evolutionary psychology and the emotions. In M. Lewis & J. M. Haviland-Jones (Eds.), *Handbook of Emotions* (2nd ed.) (pp. 91–115.) New York: Guilford; Tice, D. M., Baumeister, R. F., Shmueli, D., & Muraven, M. (2007). Restoring the self: Positive affect helps improve self-regulation following ego depletion. *Journal of Experimental Social Psychology, 43,* 379–384.

29. This sentence reflects a wide confluence of thinking in psychology. For example, Carver, C. S., & Scheier, M. F. (1990). Origins and functions of positive and negative affect: A control-process view. *Psychological Review*, *97*, 19–35; Klinger, E. (1975). Consequences of commitment to and disengagement from incentives. *Psychological Review*, *82*, 1–25.

30. Extensive experimental work of Gendolla shows that negative moods mobilize effort in the body particularly when a task becomes more difficult, but not when the task is judged to be impossible. For a review see Gendolla, G. H. E. (2000). On the impact of mood on behavior: An integrative theory and a review. *Review of General Psychology*, *4*, 378–408.

31. Heckhausen, J., Wrosch, C., & Fleeson, W. (2001). Developmental regulation before and after a developmental deadline: The sample case of "biological clock" for childbearing. *Psychology and Aging*, *16*, 400–413.

32. Nesse, R. M. (2000). Is depression an adaptation? *Archives of General Psychiatry*, *57*, 14–20.

33. Knowing that mood evolved way before language helps us understand why it's not so easy to reason one's way out of a depression!

34. Though other mammals probably have internal experiences of mood similar to humans, we cannot know this, and the present argument does not rest on this idea. Baars, B. J. (2005). Subjective experience is probably not limited to humans: The evidence from neurobiology and behavior. *Consciousness and Cognition*, *14*, 7–21.

35. Multiple lines of evidence converge on the primitivity of affect, including (a) studies that show a relationship between direct brain stimulation and affect, (b) work showing that aspects of fear can be activated without recognition of a fear stimulus, (c) work showing that bodily changes herald a change in decision making, and (d) evidence of emotions/moods in human infants. See Damasio, A. R. (1996). The somatic marker hypothesis and the possible functions of the prefrontal cortex. *Philosophical Transactions of the Royal Society of London, Series B: Biological Sciences*, *351*, 1413–1420; Damasio, A. R. (1994). *Descartes' error: Emotion, reason, and the human brain*. New York: Grosset/Putnam; Ohman, A., & Soares, J. J. F. (1994). "Unconscious anxiety": Phobic responses to masked stimuli. *Journal of Abnormal Psychology*, *103*, 231–240; LeDoux, J. E. (1996). *The emotional brain: The mysterious underpinnings of emotional life*. New York: Simon & Schuster.

36. Darwin's description of grief captures these changes nicely: "They no longer wish for action, but remain motionless and passive, or may occasionally rock themselves to and fro. The circulation becomes languid; the face pale; the muscles flaccid; the eyelids droop; the head hangs on the contracted chest; the lips, checks and lower jaw all sink downwards from their own weight." Darwin, C. R. (1872). *The expression of the emotions in man and animals* (p. 178). London: John Murray.

37. William Styron's *Darkness Visible* has a great discussion of the difficulty of putting the experience of serious depression into words.

38. Ever since William James published "What Is an Emotion?" in *Mind* in 1884 (Vol. 9, pp. 188–205), scholars have agonized over the relationships among emotion's various component parts.

39. For a great discussion of how error prone our story making about mood is, see Gilbert, D. (2006). *Stumbling on happiness.* New York: Vintage.

40. Frijda, N. H. (1993). Moods, emotion episodes, and emotions. In M. Lewis & J. M. Haviland (Eds.), *Handbook of emotions* (pp. 381–403). New York: Guilford Press.

Chapter Two: Where the Depths Begin

1. Another adaptation that presents both a good and a bad face is vomiting. On the surface, it's difficult to imagine a behavior that looks more like a manifestation of illness. We vomit when we are sick. But as Randolph Nesse and George Williams point out, vomiting is actually an adaptation. Although it does not feel pleasant, it is a potentially lifesaving defense that quickly expels toxic substances from the body. However, vomiting is not cost free. If it continues for a protracted period of time, it can contribute to dehydration and even death. Nesse, R. M., & Williams, G. C. (1996). *Why we get sick: The new science of Darwinian medicine.* New York: Vintage.

2. Gruber, J., Mauss, I. B., & Tamir, M. (2011). A dark side of happiness? How, when, and why happiness is not always good. *Perspectives on Psychological Science, 6,* 222–233.

3. See Pinker, S. (1997). *How the mind works.* New York: W. W. Norton & Company. Pinker argues, importantly, that accepting evolution as a fact does not mean we lose the right to judge it from a moral standpoint. Even if low mood were a completely perfect adaptation, we could still be critical about how much of it we wanted as a society. Otherwise we are accepting the argument that whatever is, is right.

4. "The appearance of dejection in young orangs and chimpanzees, when out of health, is as plain and almost as pathetic as in the case of our children. This state of mind and body is shown by their listless movements, fallen countenances, dull eyes, and changed complexion." Darwin, C. R. (1872). *The expression of the emotions in man and animals.* London: John Murray.

5. On de-escalating conflict, Gilbert, P. (1992). *Depression: The evolution of powerlessness.* East Sussex, UK: Lawrence Erlbaum Associates; on social risk, Allen, N. B., & Badcock, P. B. T. (2003). The social risk hypothesis of depressed mood: Evolutionary, psychosocial, and neurobiological perspectives. *Psychological Bulletin, 129,* 887–913; on braking effort: Nesse, R. M. (2000). Is depression an adaptation? *Archives of General Psychiatry, 57,* 14–20; and on complex problems, Andrews, P. W., & Thomson, J. A., Jr.

(2009).The bright side of being blue: Depression as an adaptation for analyzing complex problems. *Psychological Review, 116*, 620–654.

6. This literature on the benefits of low mood is rather human centric and only addresses some of its functions. See, for example, Ambady, N., & Gray, H. (2002). On being sad and mistaken: Mood effects on the accuracy of thin-slice judgments. *Journal of Personality and Social Psychology, 83*, 947–961; Edwards, J. A., & Weary, G. (1993). Depression and the impression-formation continuum: Piecemeal processing despite the availability of category information. *Journal of Personality and Social Psychology, 64*, 636–645; Forgas, J. P. (1998). On being happy and mistaken: Mood effects on the fundamental attribution error. *Journal of Personality and Social Psychology, 75*, 318–331; Gasper, K. (2004). Do you see what I see? Affect and visual information processing. *Cognition and Emotion, 18*, 405–421; Gasper, K., & Clore, G. L. (2002). Attending to the big picture: Mood and global versus local processing of visual information. *Psychological Science, 13*, 34–40; and Schwarz, N. (1990). Feelings as information: Informational and motivational functions of affective states. In E. T. Higgins & R. M. Sorrentino (Eds.), *Handbook of motivation and cognition: Foundations of social behavior* (Vol. 2, pp. 527–561). New York: Guilford Press.

7. Alloy, L. B., & Abramson, L. Y. (1979). Judgment of contingency in depressed and nondepressed students: Sadder but wiser? *Journal of Experimental Psychology: General, 108*, 441–485.

8. Some evidence points to the conclusion that people in a depressed mood may be more evenhanded in weighing information about themselves, whereas nondepressed people tend to hold overly positive views about themselves. Ahrens, A. H., Zeiss, A. M., & Kanfer, R. (1988). Dysphoric deficits in interpersonal standards, self-efficacy, and social comparison. *Cognitive Therapy and Research, 12*, 53–67; Alloy, L. B., & Ahrens, A. H. (1987). Depression and pessimism for the future: Biased use of statistically relevant information in predictions for self versus others. *Journal of Personality and Social Psychology, 52*, 366–378.

9. Forgas, J. P. (2007). When sad is better than happy: Negative affect can improve the quality and effectiveness of persuasive messages and social influence strategies. *Journal of Experimental Social Psychology, 43*, 513–528.

10. Experiments described in Forgas, J. P. (2013). Don't worry, be sad! On the cognitive, motivational, and interpersonal benefits of negative mood. *Current Directions in Psychological Science, 22*, 225–232.

11. Ackermann, R., & DeRubeis, R. J. (1991). Is depressive realism real? *Clinical Psychology Review, 11*, 565–584; Moore, M. T., & Fresco, D. M. (2012). Depressive realism: A meta-analytic review. *Clinical Psychology Review, 32*, 496–509.

12. Taylor, S. E., & Brown, J. D. (1988). Illusion and well-being: A social psychological perspective on mental health. *Psychological Bulletin, 103*, 193–210.

13. Wrosch C., &. Miller, G. E. (2009). Depressive symptoms can be useful: Self-regulatory and emotional benefits of dysphoric mood in adolescence. *Journal of Personality and Social Psychology, 96,* 1181–1190.

14. Bowlby, J. (1980). *Attachment and loss. Vol. 3 Loss: Sadness and depression.* New York: Basic Books.

15. Banks, S. M., & Kerns, R. D. (1996). Explaining high rates of depression in chronic pain: A diathesis-stress framework. *Psychological Bulletin, 119,* 95–110; Zuroff, D. C., Fournier, M. A., & Moskowitz, D. S. (2007). Depression, perceived inferiority, and interpersonal behavior: Evidence for the involuntary defeat strategy. *Journal of Social and Clinical Psychology, 26,* 751–778.

16. For example, Hippocrates described low mood and depression more than twenty-five hundred years ago in terms that resonate well with the present day.

17. Transient sadness may lead to greater caution when reasoning about hypothetical life scenarios. Yuen, K. S. L., & Lee, T. M. C. (2003). Could mood state affect risk-taking decisions? *Journal of Affective Disorders, 75,* 11–18; Chou, K. L., Lee, T., & Ho, A. H. Y. (2007). Does mood state change risk taking tendency in older adults? *Psychology and Aging, 22,* 310–318. Logically, people who never become sad should lead lives filled with repeated mistakes, but there is little empirical work on this issue.

18. Stringer, L. (2002). Fading to gray. In N. Casey (Ed.), *Unholy ghost: Writers on depression* (p. 113). New York: William Morrow.

19. Nesse, R. M. (2000). Is depression an adaptation? *Archives of General Psychiatry, 57,* 14–20. This is also the main thrust of the analytical rumination hypothesis.

20. Lykouras, E., Malliaras, D., Christodoulou, G. N., Papakostas, Y., Voulgari, A., Tzonou, A., & Stefanis, C. (1986). Delusional depression: Phenomenology and response to treatment. *Acta Psychiatrica Scandinavica, 73,* 324–329.

21. Frenk Peeters, personal communication with the author, December 1, 2010.

22. Thompson, T. (1996). *The beast: A journey through depression.* New York: Plume, p. 3.

23. Snyder, H. R. (2013). Major depressive disorder is associated with broad impairments on neuropsychological measures of executive function: A meta-analysis and review. *Psychological Bulletin, 139,* 81–132.

24. Although I have focused on negative mood, this analysis could easily be applied to other negative emotional states. For example, anger is unlikely to be absolutely good or bad. Anger is sometimes useful in mobilizing action against a serious offense and sometimes not useful, such as when chronic anger and poor anger management destroy close relationships. By extension, a detailed understanding of the costs and benefits of any emotional state puts us in a better position to minimize its costs and maximize its benefits.

25. For examples of the polarized debate, see Lehrer, J. (2010, February 25). Depression's upside. *New York Times*. Retrieved from http://www.ny times.com/2010/02/28/magazine/28depression-t.html?pagewanted=all& _r=0; Andrews, P. W., & Thomson, J. A., Jr. (2009). The bright side of being blue: Depression as an adaptation for analyzing complex problems. *Psychological Review, 116*, 620–654; and Pies, R. W. (2010). The myth of depression's upside [Web log post]. Retrieved from http://psychcentral.com/blog /archives/2010/03/01/the-myth-of-depressions-upside/.

26. Kramer, P. D. (2005). *Against depression*. New York: Viking.

27. Von Helversen, B., Wilke, A., Johnson, T., Schmid, G., & Klapp, B. (2011). Performance benefits of depression: sequential decision making in a healthy sample and a clinically depressed sample. *Journal of Abnormal Psychology, 120*, 962–968.

28. See Levinson, D. F. (2009). Genetics of major depression. In I. H. Gotlib & C. L. Hammen (Eds.), *Handbook of depression* (2nd ed.) (pp. 165–186). New York: Guilford Press.

29. See, for example, Ormel, J., Oldehinkel, A. J., & Brilman, E. I. (2001). The interplay and etiological continuity of neuroticism, difficulties, and life events in the etiology of major and subsyndromal, first and recurrent depressive episodes in later life. *American Journal of Psychiatry, 158*, 885–891; and Horowitz, A., Reinhardt, J. P., & Kennedy, G. J. (2005). Major and subthreshold depression among older adults seeking vision rehabilitation services. *American Journal of Geriatric Psychiatry, 13*, 180–187.

30. Iacoviello, B. M., Alloy, L. B., Abramson, L. Y., & Choi, J. Y. (2010). The early course of depression: A longitudinal investigation of prodromal symptoms and their relation to the symptomatic course of depressive episodes. *Journal of Abnormal Psychology, 119*, 459–467; Murphy, J. M., Sobol, A. M., Olivier, D. C., Monson, R. R., Leighton, A. H., & Pratt, L. A. (1989). Prodromes of depression and anxiety: The Stirling County study. *British Journal of Psychiatry, 155*, 490–495.

31. Fava, G. A., Grandi, S., Zielezny, M., Canestrari, R., & Morphy, M. A. (1994). Cognitive behavioral treatment of residual symptoms in primary major depressive disorder. *American Journal of Psychiatry, 151*, 1295–1299. These authors also showed that the nature of the residual symptoms was similar to that of the symptoms that preceded the depression.

32. Boland, R. J., & Keller, M. B. (2009). Course and outcome of depression. In I. H. Gotlib & C. L. Hammen (Eds.), *Handbook of depression* (2nd ed.) (pp. 23–43). New York: Guilford Press.

33. The median length of an episode of depression varies across studies, anywhere from four to seven months. Solomon, D. A., Keller, M. B., Leon, A. C., Mueller, T. I., Shea, M. T., Warshaw, M., . . . Endicott J. (1997). Recovery from major depression: A 10-year prospective follow-up across multiple episodes. *Archives of General Psychiatry, 54*, 1001–1006; Eaton, W. W., Anthony, J. C., Gallo, J., Cai, G., Tien, A., Romanoski, A., . . . Chen

L. S. (1997). Natural history of Diagnostic Interview Schedule/DSM-IV major depression: The Baltimore Epidemiological Catchment Area follow-up. *Archives of General Psychiatry*, *54*, 993–999; Keller, M. B., Lavori, P. W., Mueller, T. I., Endicott, J., Coryell, W., Hirschfeld, R. M., & Shea T. (1992). Time to recovery, chronicity, and levels of psychopathology in major depression. *Archives of General Psychiatry*, *49*, 809–816.

34. As with depression, the obesity epidemic results from an interaction of our genetic makeup with the modern environment. Gluckman, P. D., & Hanson, M. A. (2008). Developmental and epigenetic pathways to obesity: An evolutionary-developmental perspective. *International Journal of Obesity*, *32*, S62–S71.

35. Sapolsky, R. M. (2004). *Why zebras don't get ulcers* (3rd ed.). New York: Henry Holt and Company.

36. On the surface, deep depression appears to be more costly than shallow depression, but this does not mean that deep depression invariably lowers fitness. It may be a potentially expensive response that is helpful when deployed sparingly (as discussed in later chapters).

37. One mystery I will take up is why depression would increase during the same period that objective material and health conditions for most people living on the planet were improving; see Easterman, G. (2003). *The progress paradox*. New York: Random House. This suggests that levels of objective health and resources are not a sufficient account of mood.

38. Cosmides, L., & Tooby, J., *Evolutionary psychology: A primer*. (1997). Retrieved from http://www.psych.ucsb.edu/research/cep/primer.html.

39. Most scientists think there has been only modest change in the human genome in recent times, although there is active discussion on this point. Cochran, G., & Harpending, H. (2009). *The 10,000 year explosion: How civilization accelerated human evolution*. New York: Basic Books.

Chapter Three: What Other Species Tell Us About Depression

1. The building blocks of mood should be observed widely in the animal kingdom. For discussion of this issue, see Damasio, A. M. (1999). *The feeling of what happens: Body and emotion in the making of consciousness*. New York: Houghton Mifflin Harcourt. See also Francis Crick Memorial Conference. (2012, July 7). The Cambridge declaration on consciousness. Retrieved from http://fcmconference.org/img/CambridgeDeclarationOn Consciousness.pdf.

2. In experiments by Sarah Brosnan, chimpanzees and monkeys that traded tokens for cucumbers responded negatively once they saw that other animals were getting a tastier treat—grapes—for the same price. Angry reactions included vocalizations and throwing away their cucumbers or their tokens and seemed to signal some recognition of a missed opportunity. Tierney, J. (2009, June 1). In that tucked tail, real pangs of regret? *New York*

Times. Retrieved from http://www.nytimes.com/2009/06/02/science/02tier .html?_r=0.

3. For preschool depression epidemiology, see Egger, H. L., & Angold, A. (2006). Common emotional and behavioral disorders in preschool children: Presentation, nosology, and epidemiology. *Journal of Child Psychology and Psychiatry, 47,* 313–337; and Paul P. (2010, August 25). Can preschoolers be depressed? *New York Times.* Retrieved from http://www .nytimes.com/2010/08/29/magazine/29preschool-t.html?pagewanted=all.

4. Keltner, D., & Lerner, J. S. (2010). Emotion. In S. T. Fiske, D. T. Gilbert, & G. Lindzey (Eds.), *Handbook of social psychology* (5th ed.) (Vol.1, pp. 317–352). Hoboken, NJ: John Wiley & Sons.

5. Kleinman, A. (1988). *Rethinking psychiatry: From cultural category to personal experience.* New York: Free Press; Kleinman, A., & Good, B. (Eds.). (1985). *Culture and depression: Studies in the anthropology and cross-cultural psychiatry of affect and disorder.* Berkeley: University of California Press; Ballenger, J. C., Davidson, J. R., Lecrubier, Y., Nutt, D. J., Kirmayer, L. J., Lépine, J. P., . . . Ono, Y. (2001). Consensus statement on transcultural issues in depression and anxiety from the International Consensus Group on Depression and Anxiety. *Journal of Clinical Psychiatry, 62,* 47–55.

6. If we assume that every species has a unique ecological niche and has been adapted to fit that niche, we should expect depression in different mammals to exhibit variations on a theme (i.e., we should not expect bat depression to manifest in the exact same ways as cat depression).

7. Shively, C.A., Register, T. C., Friedman, D. P., Morgan, T. M., Thompson, J., & Lanier, T. (2005). Social stress-associated depression in adult female cynomolgus monkeys (Macaca fascicularis). *Biological Psychology, 69,* 67–84.

8. Interestingly, administering antidepressants can reverse these abnormalities: Baltzer, V., & Weiskrantz, L. (1973). Antidepressant agents and reversal of diurnal activity cycles in the rat. *Biological Psychiatry, 10,* 199–209; Willner, P. (Ed.) (1991). *Behavioural models in psychopharmacology: Theoretical, industrial and clinical perspectives.* New York: Cambridge University Press.

9. Mice who were bred to be especially prone to exhibit depressive behaviors (immobility in the tail test) showed lighter and more fragmented sleep and decreased rapid eye movement (REM) sleep latency, abnormalities resembling those observed in depressed patients. El Yacoubi, M., Bouali, S., Popa, D., Naudon, L., Leroux-Nicollet, I., Hamon, M., . . . Vaugeois, J. M. (2003). Behavioral, neurochemical, and electrophysiological characterization of a genetic mouse model of depression. *Proceedings of the National Academy of Sciences of the United States of America, 100,* 6227–6232.

10. Dog depression: Causes and cures [Web log post]. (n.d.). Retrieved from http://www.thedogdaily.com/health/illness/dog_depression/index.html ?target=depression#axzz23RJfJIFr.

11. Horowitz, A., Jacobson, D., McNichol, T., & Thomas, O., 101 dumbest moments in business 2007. (2008, January 16). *Fortune.* Retrieved from http://money.cnn.com/galleries/2007/fortune/0712/gallery.101_dumbest .fortune/2.html.

12. Crowell-Davis, S. L., & Murray, T. (2006). *Veterinary psychopharmacology.* Ames, IA: Blackwell Publishing, p. 4. "As this book goes to press, the use of most psychoactive medications in veterinary medicine is extra-label. The only label uses of psychoactive medication for the treatment of behavior problems in animals are Clomicalm (clomipramine) for separation anxiety in dogs and Anipryl (L-deprenyl) for cognitive dysfunction in elderly dogs. Extra-label use means that the medication has not been approved by the Food and Drug Administration (FDA) for the specific problem and the specific species for which it is being prescribed." For further information, see Comyn, G. (2003 March/April). Extra-label drug use in veterinary medicine. *FDA Veterinarian Newsletter, 18* (2). Retrieved from http://www.fda.gov/Animal Veterinary/NewsEvents/FDAVeterinarianNewsletter/ucm100268.htm.

13. For clinical trial data on Reconcile, see http://www.fda.gov/downloads /AnimalVeterinary/Products/ApprovedAnimalDrugProducts/FOIADrug Summaries/ucm062326.pdf.

14. Crowell-Davis, S. L., & Murray, T. (2006). *Veterinary psychopharmacology.* Ames, IA: Blackwell Publishing. Controlled data on the course of depression in common household pets (i.e., cats, dogs) are sparse. Evidence that antidepressants improve cognitive performance on a cognitive test in dogs is provided in Bruhwyler, J., Chleide, E., Rettori, M. C., Poignant, J. C., & Mercier, M. (1993). Amineptine improves the performance of dogs in a complex temporal regulation schedule. *Pharmacology Biochemistry and Behavior, 45,* 897–903.

15. Matthews, K., Christmas, D., Swan, J., & Sorrell, E. (2005). Animal models of depression: Navigating through the clinical fog. *Neuroscience and Biobehavioral Reviews, 29,* 503–513; Deussing, J. M. (2006). Animal models of depression. *Drug Discovery Today: Disease Models, 3,* 375–383; Willner, P. (1984). The validity of animal models of depression. *Psychopharmacology, 83,* 1–16.

16. Steru, L., Chermat, R., Thierry, B., & Simon, P. (1985). The tail suspension test: A new method for screening antidepressants in mice. *Psychopharmacology, 85,* 367–370.

17. A demonstration of this test in action is available at www.youtube.com /watch?v=pXqDV5nSZyA.

18. Porsolt, R. D., Le Pichon, M., & Jalfre, M. (1977). Depression: A new animal model sensitive to antidepressant treatments. *Nature, 266,* 730–732; a demonstration of this test in action is available at http://www.youtube.com /watch?v=U2ngNQFv04A.

19. Porsolt, R. D., Anton, G., Blavet, N., & Jalfre, M. (1978). Behavioural despair in rats: A new model sensitive to antidepressant treatments. *European Journal of Pharmacology, 47,* 379–391.

20. Knutson, B., Wolkowitz, O. M., Cole, S. W., Chan, T., Moore, E. A., Johnson, R. C., . . . Reus, V. I. (1998). Selective alteration of personality and social behavior by serotonergic intervention. *American Journal of Psychiatry, 155,* 373–379.

21. See, for example, Seligman, M. E. P., & Beagley, G. (1975). Learned helplessness in the rat. *Journal of Comparative and Physiological Psychology, 88,* 534–541.

22. Maier, S. F. (1984). Learned helplessness and animal models of depression. *Progress in Neuro-Psychopharmacology and Biological Psychiatry, 8,* 435–446.

23. Frank, E., Tu, X. M., Anderson, B., Reynolds, C. F., III, Karp, J. F., Mayo, A., & Kupfer, D. J. (1996). Effects of positive and negative life events on time to depression onset: An analysis of additivity and timing. *Psychological Medicine, 26,* 613–624.

24. See Katz, R. J. (1982). Animal model of depression: Pharmacological sensitivity of a hedonic deficit. *Pharmacology Biochemistry and Behavior, 16,* 965–968.

25. Willner, P., Muscat, R., & Papp, M. (1992). Chronic mild stress-induced anhedonia: A realistic animal model of depression. *Neuroscience and Biobehavioral Reviews, 16,* 525–534.

26. In fact, a rich vein of work exploits variability between animals to understand vulnerability to depression, particularly genetic work with "knockout" and transgenic mice, in which rats are bred to respond more or less strongly in these sorts of situations.

27. Rottenberg, J., Ray, R. D., & Gross, J. J. (2007). Emotion elicitation using films. In J. A. Coan & J. J. B. Allen (Eds.), *The handbook of emotion elicitation and assessment* (pp. 9–28). London: Oxford University Press.

28. Hatotani, N., Nomura, J., & Kitayama, I. (1982). Changes of brain monoamines in the animal model for depression. In S. Z. Langer, R. Takahashi, T. Segawa, & M. Briley (Eds.), *New vistas in depression* (pp. 65–72). New York: Pergamon Press.

29. Although Harlow's experiments may strike the modern reader as both pointless and heinously cruel, at the time they bucked the prevailing wisdom about what was necessary for normal emotional development and highlighted the unique importance of maternal nurturance.

30. Berton, O., McClung, C. A., DiLeone, R. J., Krishnan, V., Renthal, W., Russo, S. J., Nestler, E. J. (2006). Essential role of BDNF in the mesolimbic dopamine pathway in social defeat stress. *Science, 311,* 864–868; Tsankova, N. M., Berton, O., Renthal, W., Kumar, A., Neve, R. L., & Nestler, E. J. (2006). Sustained hippocampal chromatin regulation in a mouse model of depression and antidepressant action. *Nature Neuroscience, 9,* 519–525.

31. Kalin, N. H., & Carnes, M. (1984). Biological correlates of attachment bond disruption in humans and nonhuman primates. *Progress in Neuro-*

Psychopharmacology and Biological Psychiatry, 8, 459–469; Kaufman, I. C., & Rosenblum, L. A. (1967). The reaction to separation in infant monkeys: Anaclitic depression and conservation-withdrawal. *Psychosomatic Medicine, 29,* 648–675; McKinney, W. T., & Bunney, W. E. (1969). Animal model of depression. I. Review of evidence: Implications for research. *Archives of General Psychiatry, 21,* 240–248.

32. Suomi, S. J., Eisele, C. D., Grady, S. A., & Harlow, H. F. (1975) Depressive behavior in adult monkeys following separation from family environment. *Journal of Abnormal Psychology, 84,* 576–578; Bowden, D. M., & McKinney, W. T. (1972). Behavioral effects of peer separation, isolation, and reunion on adolescent male rhesus monkeys. *Developmental Psychobiology, 5,* 353–362.

33. These reactions are similar to those of institutionalized human children. Robertson, J., & Bowlby, J. (1952). Responses of young children to separation from their mothers. *Courier of the International Children's Centre, Paris, 2,* 131–140.

34. In fact, there was cross-fertilization and exchange of ideas between the monkey and human researchers. Van der Horst, F. C. P., LeRoy, H. A., & van der Veer, R. (2008). "When strangers meet": John Bowlby and Harry Harlow on attachment behavior. *Integrative Psychological and Behavioral Science, 42,* 370–388.

35. Bowlby, J. (1999). *Attachment: Vol. 1.* (2nd ed.). New York: Basic Books, p. 27. (Original work published 1969).

Chapter Four: The Bell Tolls: Death as a Universal Trigger

1. http://download.cell.com/current-biology/mmcs/journals/0960-9822/PIIS0960982210002186.mmc2.mpg; Biro, D., Humle, T., Koops, K., Sousa, C., Hayashi, M., & Matsuzawa, T. (2010). Chimpanzee mothers at Bossou, Guinea, carry the mummified remains of their dead infants. *Current Biology, 20,* R351–R352.

2. Dunk, M. (2008, August 19). A mother's grief: Heartbroken gorilla cradles her dead baby. *Daily Mail.* Retrieved from http://www.dailymail.co.uk/sciencetech/article-1046549/A-mothers-grief-Heartbroken-gorilla-cradles-dead-baby.html.

3. Lyons, D. M., Wang, O. J., Lindley, S. E., Levine, S., Kalin, N. H., & Schatzberg, A. F. (1999). Separation induced changes in squirrel monkey hypothalamic-pituitary-adrenal physiology resemble aspects of hypercortisolism in humans. *Psychoneuroendocrinology, 24,* 131–142; baby rhesus monkeys have increases in stress hormones with maternal separation. Levine, S., & Wiener, S. G. (1988). Psychoendocrine aspects of mother-infant relationships in nonhuman primates, *Psychoneuroendocrinology, 13,* 143–154.

4. Ritchey, R. L., & Hennessy, M. B. (1987). Cortisol and behavioral responses to separation in mother and infant guinea pigs. *Behavioral and*

Neural Biology, 48, 1–12; See also Mineka, S., & Suomi, S. J. (1978). Social separation in monkeys. *Psychological Bulletin, 85,* 1376–1400.

5. How other animal species cognitively represent death is difficult to say. However, an elaborate representation is not needed to generate the relevant behaviors. Animals can be threatened by a separation without a concept of permanent mortality. Human children under five also lack an idea of mortality as a biological concept and its finality, yet they are still distressed by death and separation. It is not clear how and when in the evolution of the human species we developed a concept of mortality that allowed for cultural and symbolic elaboration on the basic response to loss. The trappings of human bereavement such as burial are an invention of the last 100,000 years.

6. Weis, D. (2010). French poet, writer and statesman, 1790–1869. In *Everlasting wisdom* (p. 20). Rogersthorpe, UK: Paragon Publishing.

7. While intelligent commentators have made the case for bereavement-like behavior elsewhere in the animal kingdom, again, the best examples can be found among the mammals; for a fine discussion of grief in other species, see King, B. J. (2013). *How animals grieve.* Chicago: University of Chicago Press.

8. Nicely summarized in Archer, J. (1999). *The nature of grief: The evolution and psychology of reactions to loss.* New York: Routledge.

9. For high rates of depressive syndromes following bereavement, see, for example, Clayton, P. J., Halikas, J. A., & Maurice, W. L. (1972). The depression of widowhood. *British Journal of Psychiatry, 120,* 71–77; Gilewski, M. J., Farberow, N. L., Gallagher, D. E., & Thompson, L. W. (1991). Interaction of depression and bereavement on mental health in the elderly. *Psychology and Aging, 6,* 67–75; Harlow, S. D., Goldberg, E. L., & Comstock, G. W. (1991). A longitudinal study of the prevalence of depressive symptomatology in elderly widowed and married women. *Archives of General Psychiatry, 48,* 1065–1068; Futterman, A., Gallagher, D., Thompson, L. W., Lovett, S., & Gilewski, M. (1990). Retrospective assessment of martial adjustment and depression during the first 2 years of spousal bereavement. *Psychological Aging, 5,* 277–283; Zisook, S., & Shuchter, S. R. (1991). Depression through the first year after the death of a spouse. *American Journal of Psychiatry, 148,* 1346–1352; Bornstein, P. E., Clayton, P. J., Halikas, J. A., Maurice, W. L., & Robins, E. (1973). The depression of widowhood after thirteen months. *British Journal of Psychiatry, 122,* 561–566; Zisook, S., & Shuchter, S. R. (1993). Major depression associated with widowhood. *American Journal of Geriatric Psychiatry, 1,* 316–326.

10. Mojtabai, R. (2011). Bereavement-related depressive episodes: Characteristics, 3-year course, and implications for the DSM-5. *Archives of General Psychiatry, 68,* 920–928; Karam, E. G. (1994). The nosological status of bereavement-related depressions. *British Journal of Psychiatry, 165,* 48–52.

11. Researchers who study problems in bereavement use different organizing frameworks that are not tied to the diagnosis of depression (e.g., the concept of complicated grief).

12. Phrase from Greenberg, G. (2005). Misery's fogs. Is depression a diagnosis or a distraction? *Harper's Magazine, 311*, 89–94.

13. Saying that bereavement-related depressions are substantially similar to other depressions does not mean that antecedent events have no influence in how depression manifests. Some preliminary evidence indicates that different antecedents can produce slightly different symptom patterns. See, for example, Keller, M. C., & Nesse, R. M. (2006). The evolutionary significance of depressive symptoms: Different adverse situations lead to different depressive symptom patterns. *Journal of Personality and Social Psychology, 91*, 316–330; Keller, M. C., & Nesse, R. M. (2005). Is low mood an adaptation? Evidence for subtypes with symptoms that match precipitants. *Journal of Affective Disorders, 86*, 27–35.

14. For examples of controversy over this change, see commentary at http://opinionator.blogs.nytimes.com/2013/02/06/the-limits-of-psychiatry /or http://newoldage.blogs.nytimes.com/2013/01/24grief-over-new-depression -diagnosis/.

15. Zisook, S., & Kendler, K. S. (2007). Is bereavement-related depression different than non-bereavement-related depression? *Psychological Medicine, 37*, 779–794.

16. Clayton, P. J. (1975). The effect of living alone on bereavement symptoms. *American Journal of Psychiatry, 132*, 133–137; Dimond, M., Lund, D. A., & Caserta, M. S. (1987). The role of social support in the first two years of bereavement in an elderly sample. *Gerontologist, 27*, 599–604; Norris, F. H., & Murrell S. A. (1990). Social support, life events, and stress as modifiers of adjustment to bereavement by older adults. *Psychology and Aging, 5*, 429–436; Harlow, S. D., Goldberg, E. L., & Comstock, G. W. (1991). A longitudinal study of risk factors for depressive symptomatology in elderly widowed and married women. *American Journal of Epidemiology, 134*, 526–538; Nuss, W. S., & Zubenko, G. S. (1992). Correlates of persistent depressive symptoms in widows. *American Journal of Psychiatry, 149*, 346–351.

17. Clayton, P. J. (1975). The effect of living alone on bereavement symptoms. *American Journal of Psychiatry, 132*, 133–137; McHorney, C. A., & Mor, V. (1988). Predictors of bereavement depression and its health services consequences. *Medical Care, 26*, 882–893.

18. Immune changes: Linn, M. W., Linn, B. S., & Jensen, J. (1984). Stressful events, dysphoric mood, and immune responsiveness. *Psychological Reports, 54*, 219–222; Gerra, G., Monti, D., Panerai, A. E., Sacerdote, P., Anderlini, R., Avanzini, P., & Franceschi, C. (2003). Long-term immune-endocrine effects of bereavement: Relationships with anxiety levels and mood. *Psychiatry Research, 121*,145–158. Endocrine changes: Roy, A., Gallucci, W., Avgerinos, P., Linnoila, M., & Gold, P. (1988). The CRH stimulation test in bereaved subjects with and without accompanying depression. *Psychiatry Research, 25*, 145–156.

19. Karam, E. G. (1994). The nosological status of bereavement-related depressions. *British Journal of Psychiatry*, *165*, 48–52; Brent, D. A., Perper, J. A., Moritz, G., Liotus, L., Schweers, J., & Canobbio, R. (1994). Major depression or uncomplicated bereavement? A follow-up of youth exposed to suicide. *Journal of the American Academy of Child and Adolescent Psychiatry*, *33*, 231–239; Bodnar, J. C., & Kiecolt-Glaser, J. K. (1994). Caregiver depression after bereavement: Chronic stress isn't over when it's over. *Psychology and Aging*, *9*, 372–380.

20. Oakley, F., Khin, N. A., Parks, R., Bauer, L., & Sunderland, T. (2002). Improvement in activities of daily living in elderly following treatment for post-bereavement depression. *Acta Psychiatrica Scandinavica*, *105*, 231–234. Kessing, L. V., Bukh, J. D., Bock, C., Vinberg, M., & Gether, U. (2010). Does bereavement-related first episode depression differ from other kinds of first depressions? *Social Psychiatry and Psychiatric Epidemiology*, *45*, 801–808. The Kessing study compared people with bereavement related depression, non-BRD, and depression without stressful life events, finding that the three groups did not differ in responsiveness to antidepressant treatment.

21. Wakefield, J. C., Schmitz, M. F., First, M. B., & Horwitz, A. V. (2007). Extending the bereavement exclusion for major depression to other losses: Evidence from the National Comorbidity Survey. *Archives of General Psychiatry*, *64*, 433–440. There was just one difference among all the comparisons.

22. Kendler, K. S., Myers, J., & Zisook S. (2008). Does bereavement-related major depression differ from major depression associated with other stressful life events? *American Journal of Psychiatry*, *165*, 1449–1455.

23. Lebanon: Karam, E. G., Tabet, C. C., Alam, D., Shamseddeen, W., Chatila, Y., Mneimneh, Z., . . . Hamalian, M. (2009). Bereavement related and non-bereavement related depressions: A comparative field study, *Journal of Affective Disorders*, *112*, 102–110. Denmark: Kessing, L. V., Bukh, J. D., Bock, C., Vinberg, M., & Gether, U. (2010). Does bereavement-related first episode depression differ from other kinds of first depressions? *Social Psychiatry and Psychiatric Epidemiology*, *45*, 801–808. France: Corruble, E., Chouinard, V. A., Letierce, A., Gorwood, P. M., & Chouinard, G. (2009). Is DSM-IV bereavement exclusion for major depressive episode relevant to severity and pattern of symptoms? A case-control, cross-sectional study. *Journal of Clinical Psychiatry*, *70*, 1091–1097. In the French data, the patients with excludable bereavement were actually more severely depressed than "regular" depressed cases.

24. Hagar, R. (2010, November 23). Dema Guinn, a heart in sorrow. *Reno Gazette Journal*. Retrieved from http://www.rgj.com/article/2010 1123/NEWS/11210368/Dema-Guinn-heart-sorrow.

25. Brown, G. W., & Harris, T. O. (Eds.). (1989). *Life events and illness*. New York: Guilford; Mazure, C. M. (1998). Life stressors as risk factors in depression. *Clinical Psychology: Science and Practice*, *5*, 291–313.

26. Kendler, K. S., Hettema, J. M., Butera, F., Gardner, C. O., & Prescott, C. A. (2003). Life event dimensions of loss, humiliation, entrapment, and danger in the prediction of onsets of major depression and generalized anxiety. *Archives of General Psychiatry, 60,* 789–796.

27. John Grace, MD, personal communication with the author, November 13, 2010.

28. Internet memorial created for Anne Elizabeth Fullwood-Smith, October 24 1956–April 15 1997 05-3-1998 by Marcia Smith. Retrieved from http://www.virtualmemorials.com/main.php?action=view&mem_id=845&page_no=1.

29. Thompson, T. (1996). *The beast: A journey through depression.* New York: Plume, p. 73.

30. "Time heals griefs and quarrels, for we change and are no longer the same persons." Pascal, B. (1996). Penseés [1670]. In E. Ehrlich & M. De Bruhl (Eds.), *International thesaurus of quotations: Revised edition* (p. 689). New York: HarperCollins. The best sources on this statistic are Clayton, P. J. (1982). Bereavement. In E. S. Paykel (Ed.), *Handbook of affective disorders* (pp. 403–415). New York: Guilford; Zisook, S., & Shuchter, S. R. (1991). Depression through the first year after the death of a spouse. *American Journal of Psychiatry, 148,* 1346–1352. After one year, 16 percent of bereaved participants had depression; Gallagher, D. E., Breckenridge, J. N., Thompson, L. W., & Peterson, J. A. (1983). Effects of bereavement on indicators of mental health in elderly widows and widowers. *Journal of Gerontology, 38,* 565–571.

31. The work of George Bonnano has shed light on successful grief, including processes in which the bereaved are able to find hidden benefit in the loss or use it as a means to refashion relationships with the deceased. Bonanno, G. A. (2009). *The other side of sadness: What the new science of bereavement tells us about life after loss.* New York: Basic Books.

Chapter Five: The Seedbed of Low Mood

1. See Horwitz, A. V., & Wakefield, J. C. (2007). *The loss of sadness: How psychiatry transformed normal sorrow into depressive disorder.* New York: Oxford University Press.

2. These patients are much more common in primary care than are patients with major depression, with some estimates stating that they are four times as common! Barrett, J. E., Barrett, J. A., Oxman, T. E., & Gerber, P. D. (1988). The prevalence of psychiatric disorders in a primary care practice. *Archives of General Psychiatry, 45,* 1100–1106; Williams, J. W., Kerber, C. A., Mulrow, C. D., Medina, A., & Aguilar, C. (1995). Depressive disorders in primary care: Prevalence, functional disability, and identification. *Journal of General Internal Medicine, 10,* 7–12.

3. Mark T. Hegel, personal communication with the author, January 20, 2011. Minor depression does not appear to respond well to antidepressants.

With a longer follow-up, patients eventually improve with "usual care," and problem-solving treatment can expedite this process. Oxman, T. E., Hegel, M. T., Hull, J. G., & Dietrich, A. J. (2008). Problem-solving treatment and coping styles in primary care for minor depression. *Journal of Consulting and Clinical Psychology, 76,* 933–943.

4. Hegel, M. T., Oxman, T. E., Hull, J. G., Swain, K., & Swick, H. (2006). Watchful waiting for minor depression in primary care: remission rates and predictors of improvement. *General Hospital Psychiatry, 28,* 205–212.

5. Six percent of a group of people with minor depression remitted during a one-month placebo lead-in to a clinical trial. Judd, L. L., Rapaport, M. H., Yonkers, K. A., Rush, A. J., Frank, E., Thase, M. E., . . . Tollefson, G. (2004). Randomized, placebo-controlled trial of fluoxetine for acute treatment of minor depressive disorder. *American Journal of Psychiatry, 161,* 1864–1871.

6. Judd, L. L., Akiskal, H. S., & Paulus, M. P. (1997). The role and clinical significance of subsyndromal depressive symptoms (SSD) in unipolar major depressive disorder. *Journal of Affective Disorders, 45,* 5–18.

7. Judd, L. L., Akiskal, H. S., & Paulus, M. P. (1997). The role and clinical significance of subsyndromal depressive symptoms (SSD) in unipolar major depressive disorder. *Journal of Affective Disorders, 45,* 5–18; Gonzalez-Tejera, G., Canino, G., Ramirez, R., Chavez, L., Shrout, P., Bird, H., & Bauermeister, J. (2005). Examining minor and major depression in adolescents. *Journal of Child Psychology and Psychiatry, 46,* 888–899.

8. Minor depression is less well researched, but several initial studies show comparable overall disability and economic burden between major and minor depression. Judd, L. L., Paulus, M. P., Wells, K. B., & Rapaport, M. H. (1996). Socioeconomic burden of subsyndromal depressive symptoms and major depression in a sample of the general population. *American Journal of Psychiatry, 153,* 1411–1417; Cuijpers, P., Smit, F., Oostenbrink, J., de Graaf, R., ten Have, M., & Beekman, A. (2007). Economic costs of minor depression: A population-based study. *Acta Psychiatrica Scandinavica, 115,* 229–236; Broadhead, W. E., Blazer, D. G., George, L. K., & Tse, C. K. (1990). Depression, disability days, and days lost from work in a prospective epidemiologic survey. *Journal of the American Medical Association, 264,* 2524–2528. Broadhead and colleagues estimated that because of its higher prevalence, minor depression was associated with more disability days than was major depression!

9. This may be a conservative estimate. Cuijpers, P., de Graaf, R., & van Dorsselaer, S. (2004). Minor depression: Risk profiles, functional disability, health care use and risk of developing major depression. *Journal of Affective Disorders, 79,* 71–79; Fogel, J., Eaton, W. W., & Ford, D. E. (2006). Minor depression as a predictor of the first onset of major depressive disorder over a 15-year follow-up. *Acta Psychiatrica Scandinavica, 113,* 36–43.

10. Deep depression not preceded by a shallow depression appears to be quite uncommon. (Judd et al. confirmed in a longitudinal study that 1 per-

cent or less of people who have zero symptoms of depression will develop deep depression in a year's time.)

11. Bolger, N., DeLongis, A., Kessler, R. C., & Schilling, E. A. (1989). Effects of daily stress on negative mood. *Journal of Personality and Social Psychology, 57,* 808–818.

12. There is work showing that daily stressors do not, on average, appear to affect mood beyond the day of their occurrence. Rehm, L. P. (1978). Mood, pleasant events, and unpleasant events: Two pilot studies. *Journal of Consulting and Clinical Psychology, 46,* 854–859; Stone, A. A., & Neale, J. M. (1984). Effects of severe daily events on mood. *Journal of Personality and Social Psychology, 46,* 137–144.

13. One study found a contrast effect, such that high negative affect on one day was followed by low negative affect the next day. Williams, K. J., Suls, J., Alliger, G. M., Learner, S. M., & Wan, C. K. (1991). Multiple role juggling and daily mood states in working mothers: An experience sampling study. *Journal of Applied Psychology, 76,* 664–674. Similarly, a diary study that took eight measurements per day for eight days found that the previous day's mood and problems were unrelated to mood during the next day. Marco, C. A., & Suls, J. (1993). Daily stress and the trajectory of mood: Spillover, response assimilation, contrast, and chronic negative affectivity. *Journal of Personality and Social Psychology, 64,* 1053–1063. Perhaps most remarkably, in a classic study Stone and Neale found no effects of "severe" daily events on the next day's mood in a volunteer sample. What is particularly interesting about this study is that they took great pains to sift through their sample to find men who had encountered a severe negative event that influenced mood that day. The next day's mood was examined using both self-reports from the men and wives' reports about their husbands' mood. Negative events did not have a strong impact on mood the next day or on days after that. Stone, A. A., & Neale, J. M. (1984). Effects of severe daily events on mood. *Journal of Personality and Social Psychology, 46,* 137–144.

14. Daniel Gilbert's (2006) work on affective forecasting—summarized in *Stumbling on Happiness* (New York: Vintage)—shows that people's real reaction to good and bad events often lasts for a shorter amount of time than they would have predicted before the event occurred.

15. Shapiro, D., Jamner, L. D., Goldstein, I. B., & Delfino, R. J. (2001). Striking a chord: Moods, blood pressure, and heart rate in everyday life. *Psychophysiology, 38,* 197–204.

16. Zelenski, J. M., & Larsen, R. J. (2000). The distribution of basic emotions in everyday life: A state and trait perspective from experience sampling data. *Journal of Research in Personality, 34,* 178–197.

17. More than half of first episodes of depression are preceded by severe life events as defined by the investigator: the person may or may not be explicitly aware of any link between depression and the event in question. Monroe, S. M., & Harkness, K. L. (2005). Life stress, the "kindling"

hypothesis, and the recurrence of depression: Considerations from a life stress perspective. *Psychological Review, 112,* 417–445.

18. For example, Wilson, T. D., & Gilbert, D. T. (2008). Explaining away: A model of affective adaptation. *Perspectives on Psychological Science, 3,* 370–386.

19. This overlaps with the analytical rumination hypothesis, that low mood is aroused by "difficult" social dilemmas. For example, Alzheimer's caregiving is associated with chronic low mood, dealing with a group of patients who arouse frustrating and ambivalent reactions over a worsening course of illness.

20. Chapman, D. P., Whitfield, C. L., Felitti, V. J., Dube, S. R., Edwards, V. J., & Anda, R. F. (2004). Adverse childhood experiences and the risk of depressive disorders in adulthood. *Journal of Affective Disorders, 82,* 217–225. One indication that the timing matters is the comparison between the effects of child and adult traumas on depression: traumas that happen in childhood increase the chances for later mood problems more than those that happen in adulthood. Molnar, B. E., Buka, S. L., & Kessler, R. C. (2001). Child sexual abuse and subsequent psychopathology: Results from the National Comorbidity Survey. *American Journal of Public Health, 91,* 753–760; MacMillan, H. L., Fleming, J. E., Streiner, D. L., Lin, E., Boyle, M. H., Jamieson, E., . . . Beardslee, W. R. (2001). Childhood abuse and lifetime psychopathology in a community sample. *American Journal of Psychiatry, 158,* 1878–1883; Mullen, P. E., Martin, J. L., Anderson, J. C., Romans, S. E., & Herbison, G. P. (1996). The long-term impact of the physical, emotional, and sexual abuse of children: A community study. *Child Abuse and Neglect, 20,* 7–21; Kendler, K. S., Bulik, C. M., Silberg, J., Hettema, J. M., Myers, J., & Prescott, C. A. (2000). Childhood sexual abuse and adult psychiatric and substance use disorders in women: An epidemiological and cotwin control analysis. *Archives of General Psychiatry, 57,* 953–959.

21. Notably, her depressive episode followed additional provocations, such as both her parents having heart attacks, a grandparent dying, and the drama of finally sharing the secret horror of her childhood. It is possible to bump along with low mood for decades.

22. Galea, S., Ahern, J., Resnick, H., Kilpatrick, D., Bucuvalas, M., Gold, J., & Vlahov, D. (2002). Psychological sequelae of the September 11 terrorist attacks in New York City. *New England Journal of Medicine, 346,* 982–987.

23. Classic work on infant temperament by Thomas, A., & Chess, S. (1977). *Temperament and development.* New York: Brunner/Mazel.

24. Kagan, J., & Snidman, N. (1991). Temperamental factors in human development. *American Psychologist, 46,* 856–862.

25. See, for example, Posner, M. I., Rothbart, M. K., & Sheese, B. E. (2007). Attention genes. *Developmental Science, 10,* 24; Loehlin, J. C., McCrae, R. R., Costa, P. T., & John, O. P. (1998). Heritabilities of common and measure specific components of the Big Five personality factors. *Journal of Research*

in Personality, 32, 431–453. I am very much emphasizing the human case, but there is equally overwhelming evidence for a wide spectrum of temperaments in other mammals. Anyone who has been around puppies, for example, would know this.

26. Experiments described in Wilson, D. S. (2007). *Evolution for everyone: How Darwin's theory can change the way we think about our lives.* New York: Bantam Dell, pp. 106–108. See also Cain, S. (2001, June 25). Shyness: Evolutionary tactic. *New York Times.* Retrieved from http://www.nytimes.com/2011/06/26/opinion/sunday/26shyness.html?pagewanted=2.

27. There are demonstrations that effects of life events last longer in people who report being high on neuroticism—see Suls, J., & Martin, R. (2005). The daily life of the garden-variety neurotic: Reactivity, stressor exposure, mood spillover, and maladaptive coping. *Journal of Personality, 73*, 1485–1509—and in other species; see Weiss, A., King, J. E., & Perkins, L. (2006). Personality and subjective well-being in orangutans (Pongo pygmaeus and Pongo abelii). *Journal of Personality and Social Psychology, 90*, 501–511.

28. Space limitations require an abbreviated treatment of the lab studies here.

29. Scholars have tended to not look for countervailing advantages of neuroticism; an exception is the work of Gerald Matthews. Matthews, G. (2004). Neuroticism from the top down: Psychophysiology and negative emotionality. In R. Stelmack (Ed.), *On the psychobiology of personality: Essays in honor of Marvin Zuckerman* (pp. 249–266). Amsterdam: Elsevier Science. He argues from the extensive literature on anxiety and cognition—that people high in neuroticism have enhanced sensitivity to the detection of threats. This high sensitivity will confer performance advantages in environments where there are "subtle, disguised, or distant" dangers. However, people who are low in neuroticism may show superior performance in coping with immediate threats (e.g., extreme physical danger), whereas high neuroticism will impair performance in these contexts. The environments that humans occupy have ample niches for a wide range of variation on this trait. Traits closely related to neuroticism may also add to competitive drive, as nicely illustrated by tennis star Cliff Richey: "The up-side to anxiety, if there is one, is that it drives you towards success. The moderate down moods even help too . . . by motivating you to get out there and succeed in order to feel better. You learn to counteract pain with the tonic of success. When dysthymia hits you, it drives you on. It tells you that you aren't very good. You try to smother feelings of inadequacy by being creative and successful. Even when you win, that insecurity keeps telling you: 'You still aren't good enough.' It propels you into wanting to become even better." Richey, C. & H. R. (2010). *Acing Depression: A tennis champion's toughest match.* Chicago: New Chapter Press, n.p.

30. Neuroticism is associated with an increased propensity for a spectrum of psychological problems, such as depression, anxiety, and schizophrenia.

These problems are enumerated in Lahey, B. B. (2009). Public health significance of neuroticism. *American Psychologist, 64,* 241–256.

31. Bouchard, T. J., & Loehlin, J. C. (2001). Genes, evolution, and personality. *Behavior Genetics, 31,* 243–273; Nettle, D. (2006). The evolution of personality variation in humans and other animals. *American Psychologist, 61,* 622–631. Nettle specifically theorizes that the range of observed neuroticism in humans (and other species) represents a tradeoff between advantages (vigilance to dangers and enhanced goal striving) and costs (depression/poorer health).

32. Other factors that contribute to the escalation of low mood, such as poor mood regulation (e.g., getting drunk when you have low mood) and unrealistic attitudes and expectations about happiness, are discussed in the next chapter. Mauss, I. B., Tamir, M., Anderson, C. L., & Savino, N. S. (2011). Can seeking happiness make people unhappy? Paradoxical effects of valuing happiness. *Emotion, 11,* 807–815.

33. For evidence of a daily diurnal rhythm of positive affect, see Clark, L. A., Watson, D., & Leeka, J. (1989). Diurnal variation in the positive affects. *Motivation and Emotion, 13,* 205–234.

34. Espiritu, R. C., Kripke, D. F., Ancoli-Israel, S., Mowen, M. A., Mason, W. J., Fell, R. L., & Kaplan, O. J. (1994). Low illumination experienced by San Diego adults: Association with atypical depressive symptoms. *Biological Psychiatry, 35,* 403–407.

35. Modern air travel across time zones is another good example of how modern routines trample endogenous biological rhythms, with negative consequences for mood, as seen in jet lag.

36. National Sleep Foundation. (2011, March 7). Annual Sleep in America poll explores connections with communications technology use and sleep [Press release]. Retrieved from http://www.sleepfoundation.org/article/press-release/annual-sleep-america-poll-exploring-connections-communications-technology-use. Electronics use such as texting in the hour before bed was predictive of not feeling refreshed by a night's sleep.

37. A growing appreciation of the effects of sleep restriction, both adverse physical and mental health effects, is evident in Wiebe, S. T., Cassoff, J., & Gruber, R. (2012). Sleep patterns and the risk for unipolar depression: A review. *Nature and Science of Sleep, 4,* 63–71; and Banks, S., & Dinges, D. F. (2007). Behavioral and physiological consequences of sleep restriction. *Journal of Clinical Sleep Medicine, 3,* 519–528. A nice experimental study showing the cognitive costs of sleep restriction is Van Dongen, H. P., Maislin, G., Mullington, J. M., & Dinges, D. F. (2003). The cumulative cost of additional wakefulness: dose-response effects on neurobehavioral functions and sleep physiology from chronic sleep restriction and total sleep deprivation. *Sleep, 26,* 117–126.

38. Figures from 2011 Sleep in America Poll from the National Sleep Foundation; Bonnet, M. H., & Arand, D. L. (1995). We are chronically sleep deprived. *Sleep, 18,* 908–911.

39. Wilson, J. F. (2005). Is sleep the new vital sign? *Annals of Internal Medicine, 142,* 877–880. There is some debate about how dramatic the increase in the number of short sleepers is. National Sleep Foundation. (2005). 2005 Sleep in America poll: Summary of findings. Retrieved from http://www.sleepfoundation.org/sites/default/files/2005_summary_of_findings.pdf; Knutson, K. L., Van Cauter, E., Rathouz, P. J., DeLeire, T., & Lauderdale, D. S. (2010). Trends in the prevalence of short sleepers in the USA: 1975–2006. *Sleep, 33,* 37–45. The latter study finds the trend most pronounced among those who have full-time jobs.

40. Interestingly, in the Sleep in America poll more than one-quarter of respondents agreed with the idea that their daily routine does not allow for adequate sleep. Of those who said their routine did not allow for adequate sleep, 85 percent said that their mood was adversely impacted.

41. Judd, L. L., Akiskal, H. S., & Paulus, M. P. (1997). The role and clinical significance of subsyndromal depressive symptoms (SSD) in unipolar major depressive disorder. *Journal of Affective Disorders, 45,* 5–18.

Chapter Six: The Slide

1. See, for example, Easterbrook, G. (2004). *The progress paradox: How life gets better while people feel worse.* New York: Random House Trade Paperbacks; Hidaka, B. H. (2012). Depression as a disease of modernity: Explanations for increasing prevalence. *Journal of Affective Disorders, 140,* 205–214.

2. Given the impossibility of doing epidemiological studies with other species, it's difficult to prove the idea that depression is more common, or worse for humans than it is for other species; still, it's worth noting that we are the only species whose members regularly commit suicide when depressed.

3. There is considerable overlap between feeling bad about bad past outcomes (rumination) and feeling bad about bad outcomes in the future (worry). We should expect worry and rumination to have both costs and benefits. Davey, G. C. L., Hampton, J., Farrell, J., & Davidson, S. (1992). Some characteristics of worrying: Evidence for worrying and anxiety as separate constructs. *Personality and Individual Differences, 13,* 133–147; Siddique, H. I., LaSalle-Ricci, V. H., Glass, C. R., Arnkoff, D. B., & Diaz, R. J. (2006). Worry, optimism, and expectations as predictors of anxiety and performance in the first year of law school. *Cognitive Therapy and Research, 30,* 667–676.

4. *Psychotherapy* is a broad term that covers a variety of techniques. Most germane here is that insight-oriented psychotherapy is a type of expert-guided talk therapy that typically focuses on discovering the hidden meanings that might underpin depression. Shedler, J. (2010). The efficacy of psychodynamic psychotherapy. *American Psychologist, 65,* 98–109.

5. You might find it ironic that a scientist who studies mood would ever question the value of thinking about moods.

6. Major studies of rumination include Nolen-Hoeksema, S., Morrow, J., & Fredrickson, B. L. (1993). Response styles and the duration of episodes of depressed mood. *Journal of Abnormal Psychology, 102,* 20–28; Nolen-Hoeksema, S., & Morrow, J. (1993). Effects of rumination and distraction on naturally occurring depressed moods. *Cognition and Emotion, 7,* 561–570; Lyubomirsky, S., & Nolen-Hoeksema, S. (1995). Effects of self-focused rumination on negative thinking and interpersonal problem solving. *Journal of Personality and Social Psychology, 69,* 176–190; Nolen-Hoeksema, S., Parker, L. E., & Larson, J. (1994). Ruminative coping with depressed mood following loss. *Journal of Personality and Social Psychology, 67,* 92–104; and Nolen-Hoeksema, S., & Morrow, J. (1991). A prospective study of depression and posttraumatic stress symptoms after a natural disaster: The 1989 Loma Prieta earthquake. *Journal of Personality and Social Psychology, 61,* 115–121.

7. Scholarly work shows that depression is associated with increased self-focus, which is reflected in how depressed people describe their situations: "One of the very worst features of depression is the impossibility of communicating the reality you are in to anyone outside it. Depression is a state of absolute isolation. Every depressive is an island, or so it seems. If you could get off the island, you wouldn't be depressed, but depressed people cannot leave the island. The best you can hope for is to send up a smoke signal, if you have anything to burn and if there's anyone close enough to see it." Allan, C. (2008, November 4). Thanks Iceland, Sarah Palin and VW: You're a ray of light. *The Guardian.* Retrieved from http://www.guardian.co.uk/society/2008/nov/05/depression-mental-health.

8. See, for example, Pyszczynski, T. A., & Greenberg, J. (1992). *Hanging on and letting go: Understanding the onset, progression, and remission of depression.* New York: Springer-Verlag Publishing, pp. xi, 169; Armey, M. F., Fresco, D. M., Moore, M. T., Mennin, D. S., Turk, C. L., Heimberg, R. G., & Alloy, L. B. (2009). Brooding and pondering: Isolating the active ingredients of depressive rumination with exploratory factor analysis and structural equation modeling. *Assessment, 16,* 315–327.

9. Behaviorally focused therapy by D'Zurilla and Goldfried focuses on problem solving. D'Zurilla, T. J., & Goldfried, M. R. (1971). Problem solving and behavior modification. *Journal of Abnormal Psychology, 78,* 107–126. The consequences of rumination are worse for those who start in a more severe initial depressed mood than for those who are in only a milder depressed mood. Nolan, S. A., Roberts, J. E., & Gotlib, I. H. (1998). Neuroticism and ruminative response style as predictors of change in depressive symptomatology. *Cognitive Therapy and Research, 22,* 445–455.

10. On the contrary, work by Wenzlaff and Wegner shows that a depressed mood makes it more difficult to suppress unwanted negative thoughts. Wenzlaff, R. M., Wegner, D. M., & Roper, D. W. (1988). Depression and mental control: The resurgence of unwanted negative thoughts. *Journal of Personality and Social Psychology, 55,* 882–892.

11. Depressed people tend to avoid things; depressive avoidance may be seen in part as the rejection of all other possible behavioral outlets.

12. For discussion of cybernetic/control theories of mood, see Carver, C. S., & Scheier, M. F. (1990). Origins and functions of positive and negative affect: A control-process view. *Psychological Review, 97,* 19–35.

13. Klinger, E. (1975). Consequences of commitment to and disengagement from incentives. *Psychological Review, 82,* 1–25. This is termed the *incentive-disengagement cycle.*

14. In real life, people do not have a single response to low mood; they oscillate or shift between responses. The same person might alternately be obsessed with the meaning of the mood and attempt to "suppress" the mood by ignoring, denying, or overriding its existence. There are different ways the mood signal might be discounted, such as interpreting it solely as a bodily complaint (like a headache), and people often try to override a low mood with heavy use of drugs or alcohol.

15. See Klinger, E. (1975). Consequences of commitment to and disengagement from incentives. *Psychological Review, 82,* 1–25.

16. From the Monitoring the Future Study, http://monitoringthefuture .org/datavolumes/2006/2006dv.pdf. Bachman, J. G., Johnston, L. D., & O'Malley, P. M. (2008). *Monitoring the future: Questionnaire responses from the nation's high school seniors, 2006.* Ann Arbor: Institute for Social Research, The University of Michigan; Bachman, J. G., Johnston, L. D., & O'Malley, P. M. (2007). *Monitoring the future: A continuing study of the lifestyles and values of youth, 1976.* ICPSR07927 Vol. 4. Ann Arbor: Interuniversity Consortium for Political and Social Research [distributor]. Retrieved from http://www.icpsr.umich.edu/icpsrweb/ICPSR/studies/7927.

17. Reynolds, J., Stewart, M., MacDonald, R., & Sischo, L. (2006). Have adolescents become too ambitious? High school seniors' educational and occupational plans, 1976 to 2000. *Social Problems, 53,* 186–206; Nesse, R. M. (2004). Natural selection and the elusiveness of happiness. *Philosophical Transactions of the Royal Society of London, Series B: Biological Sciences, 359,* 1333–1347. Nesse has hypothesized that modern life has changed the structure of goals; we are more likely to face larger goals that are more difficult to relinquish and require more effort over a longer period of time. Nesse contrasts the relative ease of giving up on looking for nuts after several days of fruitless foraging with "giving up on a [PhD] programme after 5 years or a marriage after 10 years."

18. Halpern, J. (2007). *Fame junkies: The hidden truths behind America's favorite addiction.* New York: Houghton Mifflin.

19. American Society for Aesthetic Plastic Surgery. (n.d.). Cosmetic surgery national data bank statistics, 1997–2007. Retrieved from http://www .surgery.org/sites/default/files/2007stats.pdf.

20. There is evidence that, but also controversy about, whether the population is becoming more narcissistic. Narcissism involves self-concern and

the setting of highly ambitious extrinsic goals. In a series of studies, Jean Twenge and her colleagues argue that there have been increases in narcissism over time based on analyses of trends in how college students report on the Narcissism Personality Inventory. Twenge, J. M., Konrath, S., Foster, J. D., Campbell, K. W., & Bushman, B. J. (2008). Egos inflating over time: A cross-temporal meta-analysis of the Narcissistic Personality Inventory. *Journal of Personality, 76,* 875–902. But see also challenges: Trzesniewski, K. H., Donnellan, M. B., & Robins, R. W. (2008). Is "Generation Me" really more narcissistic than previous generations? *Journal of Personality, 76,* 903–917.

21. Kasser and Ryan are most closely associated with the finding that holding extrinsic goals is associated with lower well-being than holding intrinsic goals. Kasser, T., & Ryan, R. M. (1993). A dark side of the American dream: Correlates of financial success as a central life aspiration. *Journal of Personality and Social Psychology, 65,* 410–422; Kasser, T., & Ryan, R. M. (1996). Further examining the American dream: Differential correlates of intrinsic and extrinsic goals. *Personality and Social Psychology Bulletin, 22,* 280–287.

22. Examples of research examining the link between perfectionism and depression are LaPointe, K. A., & Crandell, C. J. (1980). Relationship of irrational beliefs to self-reported depression. *Cognitive Therapy and Research, 4,* 247–250; and Golin, S., & Terrell, F. (1977). Motivational and associative aspects of mild depression in skill and chance tasks. *Journal of Abnormal Psychology, 86,* 389–401.

23. Edward Watkins has argued that abstract goals are more likely to trigger rumination. Watkins, E. (2011). Dysregulation in level of goal and action identification across psychological disorders. *Clinical Psychology Review, 31,* 260–278.

24. Diener, E., & Biswas-Diener, R. (2008). *The science of optimal happiness.* Boston: Blackwell Publishing; Eid, M., & Larsen, R. J. (Eds.). (2008). *The science of subjective well-being.* New York: Guilford Press; Gilbert, D. (2006). *Stumbling on happiness.* New York: Alfred A. Knopf; Lyubomirsky, S. (2008). *The how of happiness: A scientific approach to getting the life you want.* New York: Penguin Press; Seligman, M. E. P., & Csikszentmihalyi, M. (2000). Positive psychology: An introduction. *American Psychologist, 55,* 5–14. For a critique, see Ehrenreich, B. (2009). *Bright-sided: How the relentless promotion of positive thinking has undermined America.* New York: Henry Holt and Company.

25. Eid, M., & Diener, E. (2001). Norms for experiencing emotions in different cultures: Inter- and intranational differences. *Journal of Personality and Social Psychology, 81,* 869–885.

26. Tsai, Knutson, and Fung found that European American college students placed greater value on high arousal positive affective states and less value on low arousal positive affective states than did Hong Kong Chinese

students. Tsai, J. L., Knutson, B., & Fung, H. H. (2006). Cultural variation in affect valuation. *Journal of Personality and Social Psychology, 90,* 288–307.

27. Tsai, J. L., & Wong, Y. (2007). *Socialization of ideal affect through magazines.* Unpublished manuscript. See also Tsai, J. L. (2007). Ideal affect: Cultural causes and behavioral consequences. *Perspectives on Psychological Science, 2,* 242–259.

28. Mauss, I. B., Tamir, M., Anderson, C. L., & Savino, N. S. (2011). Can seeking happiness make people unhappy? Paradoxical effects of valuing happiness. *Emotion, 11,* 807–815.

29. For more on this theme, see Nesse, R. M. (2004). Natural selection and the elusiveness of happiness. *Philosophical Transactions of the Royal Society of London, Series B: Biological Sciences, 359,* 1333–1347. See also a summary review of the "costs of pursuing happiness" by Gruber, J., Mauss, I. B., & Tamir, M. (2011). A dark side of happiness? How, when, and why happiness is not always good. *Perspectives on Psychological Science, 6,* 222–233.

30. Roemer, L., Salters, K., Raffa, S. D., & Orsillo, S. M. (2005). Fear and avoidance of internal experiences in GAD: Preliminary tests of a conceptual model. *Cognitive Therapy and Research, 29,* 71–88; Hayes, S. C., Strosahl, K., &Wilson, K. G. (1999). *Acceptance and commitment therapy: An experiential approach to behavior change.* New York: Guilford Press.

31. Shallcross, A. J., Troy, A. S., Boland, M., & Mauss, I. B. (2010). Let it be: Accepting negative emotional experiences predicts decreased negative affect and depressive symptoms. *Behaviour Research and Therapy, 48,* 921–929. Several therapies are premised on the idea of accepting negative emotional states. For a great summary and popularization, see Harris, R. (2008). *The happiness trap: How to stop struggling and start living.* Boston: Shambala Publications.

32. For an excellent discussion, see Lyubomirsky, S. (2013). *The myths of happiness.* New York: Penguin Press.

Chapter Seven: The Black Hole:
The Psychobiology of Deep Depression

1. Smith, J. (1999). *Where the roots reach for water: A personal and natural history of melancholia.* New York: North Point Press, pp. 6–7.

2. See Scrimshaw, N. S. (1987). The phenomenon of famine. *Annual Review of Nutrition, 7,* 1–21. For a more modern discussion about the consequences of prolonged food deprivation, see Keys, A., Brozek, J., Henschel, A., Mickelsen, O., & Taylor, H. L. (1950). *The biology of human starvation* (2 Vols.). Oxford: University of Minnesota Press. The experiment is described in Tucker, T. (2006). *The great starvation experiment: The heroic men who starved so that millions could live.* New York: Free Press.

3. It is a mistake to assume that an adaptation always means optimal functioning in the sense of jumping higher, running faster, or seeing farther.

4. One important question is whether there might be subtypes of deep depression that are tailored to the needs of more specific situations. The history of research on subtypes of depression is an agonized one, and the viability of subtypes remains an open question.

5. Surprisingly few theories explicitly address how moods and emotions interact. A good example is Rosenberg, E. L. (1998). Levels of analysis and the organization of affect. *Review of General Psychology, 2,* 247–270.

6. My main graduate school advisor, Ian Gotlib, was steeped in this negative information processing tradition, and he was a master of finding ways to test abstract cognitive theories with carefully controlled experiments. He originally expected our experiments to show emotional hyperreactivity in depression; he showed remarkable open-mindedness to follow the data where they led.

7. Rottenberg, J., Gross, J. J., Wilhelm, F. H., Najmi, S., & Gotlib, I. H. (2002). Crying threshold and intensity in major depressive disorder. *Journal of Abnormal Psychology, 111,* 302–312.

8. Rottenberg, J., Gross, J. J., & Gotlib, I. H. (2005). Emotion context insensitivity in major depressive disorder. *Journal of Abnormal Psychology, 114,* 627–639.

9. Bylsma, L. M., Morris, B. H., & Rottenberg, J. (2008). A meta-analysis of emotional reactivity in major depressive disorder. *Clinical Psychology Review, 28,* 676–691.

10. When I first published these papers, I expected fiery condemnation. It did not materialize.

11. Quote from William Styron in Brody, J. E. (1997, December 30). Personal health: Despite the despair of depression, few men seek treatment. *New York Times.* Retrieved from http://www.nytimes.com/1997/12/30 /science/personal-health-despite-the-despair-of-depression-few-men-seek -treatment.html.

12. Trémeau, F., Malaspina, D., Duval, F., Corrêa, H., Hager-Budny, M., Coin-Bariou, L., & Gorman, J. M. (2005). Facial expressiveness in patients with schizophrenia compared to depressed patients and nonpatient comparison subjects. *American Journal of Psychiatry, 162,* 92–101.

13. Kuppens, P., Allen, N. B., & Sheeber, L. B. (2010). Emotional inertia and psychological maladjustment. *Psychological Science, 21,* 984–991.

14. Abramson, L. Y., Metalsky, G. I., & Alloy, L. B. (1989). Hopelessness depression: A theory-based subtype of depression. *Psychological Review, 96,* 358–372. Recent extensions suggest that in addition to the negative content of attributions, depressed people tend to fixedly deploy the same attributions across different situations (explanatory inflexibility), which is also associated with poorer coping behavior. Moore, M. T., & Fresco, D. M. (2007). The relationship of explanatory flexibility to explanatory style. *Behavior Therapy, 38,* 325–332; Fresco, D. M., Williams, N. L., & Nugent, N. R. (2006). Flexibility and negative affect: Examining the associations of

explanatory flexibility and coping flexibility to each other and to depression and anxiety. *Cognitive Therapy and Research, 30,* 201–210.

15. See, for example, Drevets, W. C., Price, J. L., & Furey, M. L. (2008). Brain structural and functional abnormalities in mood disorders: Implications for neurocircuitry models of depression. *Brain Structure and Function, 213,* 93–118. This inflexibility is also frequently seen in reflexes, such as the startle response, which are ultimately tied to brain function. Allen, N. B., Trinder, J., & Brennan, C. (1999). Affective startle modulation in clinical depression: Preliminary findings. *Biological Psychiatry, 46,* 542–550.

16. Burke, H. M., Davis, M. C., Otte, C., & Mohr, D. C. (2005). Depression and cortisol responses to psychological stress: A meta-analysis. *Psychoneuroendocrinology, 30,* 846–856.

17. In severe depression, production of stress hormones goes into overdrive. Twenty-four-hour cortisol production is higher, and the switch that regulates cortisol appears to be stuck (in the on position). Varghese, F. P., & Brown, E. S. (2001). The hypothalamic-pituitary-adrenal axis in major depressive disorder: A brief primer for primary care physicians. *Primary Care Companion to the Journal of Clinical Psychiatry, 3,* 151–155.

18. Establishing how long depressions typically last is not as simple as it sounds; results depend in part on the criteria that are used to define "better," the population studied, and whether the sample is studied in the context of treatment. There is a large body of literature on this question. Posternak, M. A., Solomon, D. A., Leon, A. C., Mueller, T. I., Shea, M. T., Endicott, J., & Keller, M. B. (2006). The naturalistic course of unipolar major depression in the absence of somatic therapy. *Journal of Nervous and Mental Disease, 194,* 324–329; Angst, J. (1986). The course of affective disorders. *Psychopathology, 19,* 47–52; Hohman, L. B. (1937). A review of one hundred and forty-four cases of affective disorders—after seven years. *American Journal of Psychiatry, 94,* 303–308; Huston, P. E., & Locher, L. M. (1948). Manic-depressive psychosis: Course when treated and untreated with electric shock. *Archives of Neurology and Psychiatry, 60,* 37–48; Rennie, T. A. C., & Fowler, J. B. (1942). Prognosis in manic-depressive psychoses. *American Journal of Psychiatry, 98,* 801–814; Shobe, F. O., & Brion, P. (1971). Long-term prognosis in manic-depressive illness: A follow-up investigation of 111 patients. *Archives of General Psychiatry, 24,* 334–337. More recent studies continue to find variation in episode length, with some showing high rates of chronicity and low rates of sustained remission. Rush, A. J., Trivedi, M., Carmody, T. J., Biggs, M. M., Shores-Wilson, K., Ibrahim, H., & Crismon, M. L. (2004). One-year clinical outcomes of depressed public sector outpatients: A benchmark for subsequent studies. *Biological Psychiatry, 56,* 46–53. Others show average episode lengths of three or four months. Eaton, W. W., Shao, H., Nestadt, G., Lee, B. H., Bienvenu, O. J., & Zandi, P. (2008). Population-based study of first onset and chronicity in major depressive disorder. *Archives of General Psychiatry, 65,* 513–520; Kessler, R. C., Berglund, P.,

Demler, O., Jin, R., Koretz, D., Merikangas, K. R., . . . & Wang, P. S. (2003). The epidemiology of major depressive disorder: Results from the National Comorbidity Survey Replication (NCS-R). *Journal of the American Medical Association, 289,* 3095–3105.

19. Rottenberg, J., Kasch, K. L., Gross, J. J., & Gotlib, I. H. (2002). Sadness and amusement reactivity differentially predict concurrent and prospective functioning in major depressive disorder. *Emotion, 2,* 135–146; for a neural study, see Canli, T., Cooney, R. E., Goldin, P., Shah, M., Sivers, H., Thomason, M. E., & Gotlib, I. H. (2005). Amygdala reactivity to emotional faces predicts improvement in major depression. *Neuroreport, 16,* 1267–1270.

20. Kuppens, P., Sheeber, L. B., Yap, M. B. H., Whittle, S., Simmons, J. G., & Allen, N. B. (2012). Emotional inertia prospectively predicts the onset of depressive disorder in adolescence. *Emotion, 12,* 283–289.

21. Morris, B. H., Bylsma, L. M., & Rottenberg, J. (2009). Does emotion predict the course of major depressive disorder? A review of prospective studies. *British Journal of Clinical Psychology, 48,* 255–273.

22. Peeters, F., Berkhof, J., Rottenberg, J., & Nicolson, N. A. (2010). Ambulatory emotional reactivity to negative daily life events predicts remission from major depressive disorder. *Behaviour Research and Therapy, 48,* 754–760.

23. Tucker, T. (2006). *The great starvation experiment: The heroic men who starved so that millions could live.* New York: Free Press, Ch. 7.

24. Cacioppo, J. T., & Berntson, G. G. (1999). The affect system: Architecture and operating characteristics. *Current Directions in Psychological Science, 8,* 133–137.

25. Biology is not the only domain in which collateral damage occurs; for example, deep depression, when protracted, is likely to cause collateral damage to important social relationships.

26. Schiepers, O. J., Wichers, M. C., & Maes, M. (2005). Cytokines and major depression. *Progress in Neuro-Psychopharmacology and Biological Psychiatry, 29,* 201–217; Raison, C. L., Capuron, L., & Miller, A. H. (2006). Cytokines sing the blues: inflammation and the pathogenesis of depression. *Trends in Immunology, 27*(1), 24–31.

27. Hammen, C. (1991). Generation of stress in the course of unipolar depression. *Journal of Abnormal Psychology, 100,* 555–561; Hammen, C. (2006). Stress generation in depression: Reflections on origins, research, and future directions. *Journal of Clinical Psychology, 62,* 1065–1082.

28. Landers, A. (1993, December 5). Help for those with depression. *The Ledger,* p. 10C.

Chapter Eight: An Up and Down Thing: Improvement in Depression

1. Keller, M. B., Lavori, P. W., Mueller, T. I., Endicott, J., Coryell, W., Hirschfeld, R. M., & Shea, T. (1992). Time to recovery, chronicity, and levels

of psychopathology in major depression: A 5-year prospective follow-up of 431 subjects. *Archives of General Psychiatry, 49,* 809–816. Appropriately, these investigators had strict recovery criteria: a subject had to have eight straight weeks of no or minimal symptoms. Similar rates were found in Van Londen, L., Molenaar, R. P. G., Goekoop, J. G., Zwinderman, A. H., & Rooijmans, H. G. M. (1998). Three- to 5-year prospective follow-up of outcome in major depression. *Psychological Medicine, 28,* 731–735. Generally, patients who are seen in psychiatric settings have lower rates of full remission. For example, the Collaborative Depression Study followed patients for eighteen months and found that only about one-quarter fully remitted and stayed well over the year and a half. Shea, M. T., Elkin, I., Imber, S. D., Sotsky, S. M., Watkins, J. T., Collins, J. F., & Parloff, M. B. (1992). Course of depressive symptoms over follow-up: Findings from the National Institute of Mental Health Treatment of Depression Collaborative Research Program. *Archives of General Psychiatry, 49,* 782–787.

2. The outcome literature is voluminous and confusing. Rounsaville, B. J., Prusoff, B. A., & Padian, N. (1980). The course of nonbipolar, primary major depression: A prospective 16-month study of ambulatory patients. *Journal of Nervous and Mental Disease, 168,* 406–411. Five-year follow-ups are rare, and studies are not uniform in their remission criteria or what kind of remission is being assessed (being remitted at the time of the follow-up assessment, ever remitting during the follow-up period, etc.). Mynor-Wallis and colleagues in the United Kingdom found higher proportions (56–66 percent) of complete remission at one-year follow-up in a non-naturalistic study of a primary care setting. Mynor-Wallis, L. M., Gath, D. H., & Baker, F. (2000) Randomized controlled trial of problem-solving treatment: Antidepressant medication and combined treatment for major depression in primary care. *British Medical Journal, 320,* 26–30. Simon found 45 percent remission from a six-month follow-up period in a primary care setting. Simon, G. E. (2000). Long-term prognosis of depression in primary care. *Bulletin of the World Health Organization, 78,* 439–445. One of the largest and most intense recent treatment outcome studies was the STAR*D study (more than three thousand patients), which estimated the cumulative remission rate for patients who went through all the treatments provided (up to four) at slightly more than two-thirds. Rush, A. J., Madhukar, H., Trivedi, M. H., Wisniewski, S. R., Nierenberg, A. A., Stewart, J. W., . . . Fava, M. (2006). Acute and longer-term outcomes in depressed outpatients requiring one or several treatment steps: A STAR*D report. *American Journal of Psychiatry, 163,* 1905–1917. Posternak and colleagues estimated in the Collaborative Depression Study sample that an episode of depression had a median length of twenty-three weeks before it remitted. Posternak, M. A., Solomon, D. A., Leon, A. C., Mueller, T. I., Shea, M. T., Endicott, J., & Keller, M. B. (2006). The naturalistic course of unipolar major depression in the absence of somatic therapy. *Journal of Nervous and Mental Disease, 194,*

324–329. For epidemiological samples, see McLeod, J. D., Kessler, R. C., & Landis, K. R. (1992). Speed of recovery from major depressive episodes in a community sample of married men and women. *Journal of Abnormal Psychology, 101,* 277–286; Kendler, K. S., Walters, E. E., & Kessler, R. C. (1997). The prediction of length of major depressive episodes: Results from an epidemiological sample of female twins. *Psychological Medicine, 27,* 107–117.

3. Statement that depression is a self-limiting disorder in Hollon, S. D. (2010), Cognitive and behavior therapy in the treatment and prevention of depression. *Depression and Anxiety, 28,* 263–266; Szabadi, E., & Bradshaw, C. M. (2004). Affective disorders: 1. Antidepressants. In D. J. King (Ed.), *Seminars in clinical psychopharmacology* (2nd ed.) (pp. 178–243). London: Gaskell.

4. There is unlikely to be a single responsible self-limiting process.

5. Tadić, A., Helmreich, I., Mergl, R., Hautzinger, M., Kohnen, R., Henkel, V., & Hegerl, U. (2010). Early improvement is a predictor of treatment outcome in patients with mild major, minor or subsyndromal depression. *Journal of Affective Disorders, 120,* 86–93.

6. Szegedi, A., Jansen, W. T., van Willigenburg, A. P. P., van der Meulen, E., Stassen, H. H., & Thase, M. E. (2009). Early improvement in the first 2 weeks as a predictor of treatment outcome in patients with major depressive disorder: a meta-analysis including 6562 patients. *Journal of Clinical Psychiatry, 70,* 344–353.

7. Several randomized or naturalistic studies as well as meta-analyses have shown that early improvement after one or two weeks of treatment strongly predicts later treatment outcome. Henkel, V., Seemuller, F., Obermeier, M., Adli, M., Bauer, M., Mundt, C., . . . Riedel, M. (2009). Does early improvement triggered by antidepressants predict response/remission? Analysis of data from a naturalistic study on a large sample of inpatients with major depression. *Journal of Affective Disorders, 115,* 439–449; Stassen, H. H., Angst, J., Hell, D., Scharfetter, C., & Szegedi, A. (2007). Is there a common resilience mechanism underlying antidepressant drug response? Evidence from 2848 patients. *Journal of Clinical Psychiatry, 68,* 1195–1205; Szegedi, A., Jansen, W. T., Willigenburg, A. P., van der, M. E., Stassen, H. H., & Thase, M. E. (2009). Early improvement in the first 2 weeks as a predictor of treatment outcome in patients with major depressive disorder: a meta-analysis including 6562 patients. *Journal of Clinical Psychiatry, 70,* 344–353; Posternak, M. A., & Zimmerman, M. (2005). Is there a delay in the antidepressant effect? A meta-analysis. *Journal of Clinical Psychiatry, 66,* 148–158.

8. Stassen, H. H., Delini-Stula, A., & Angst, J. (1993). Time course of improvement under antidepressant treatment: a survival-analytical approach. *European Neuropsychopharmacology, 3,* 127–135.

9. Tang, T. Z., & DeRubeis, R. J. (1999). Sudden gains and critical sessions in cognitive-behavioral therapy for depression. *Journal of Consulting and Clinical Psychology, 67,* 894–904. Opinion is divided about whether the rapid improvements should be attributed to anything specific to this type of

treatment. Ilardi, S. S., & Craighead, W. E. (1994). The role of nonspecific factors in cognitive-behavior therapy for depression. *Clinical Psychology: Science and Practice, 1*, 138–156.

10. Tadić, A., Helmreich, I., Mergl, R., Hautzinger, M., Kohnen, R., Henkel, V., & Hegerl, U. (2010). Early improvement is a predictor of treatment outcome in patients with mild major, minor or subsyndromal depression. *Journal of Affective Disorders, 120*, 86–93. One-third or more of patients will show a robust short-term response to a placebo pill.

11. Kelly, M. A. R., Roberts, J. E., & Bottonari, K. A. (2007). Non-treatment related sudden gains in depression: The role of self-evaluation. *Behaviour Research and Therapy, 45*, 737–747. Interestingly, in this study, the gains did not hold over time.

12. Posternak, M. A., & Miller, I. (2001). Untreated short-term course of major depression: A meta-analysis of outcomes from studies using wait-list control groups. *Journal of Affective Disorders, 66*, 139–146.

13. Importantly, Parker and Blignault found that very early improvement (as soon as six days) in people not receiving treatment also predicted outcome at six and twenty weeks. Parker, G., & Blignault, I. (1985). Psychosocial predictors of outcome in subjects with untreated depressive disorder. *Journal of Affective Disorders, 8*, 73–81; Parker, G., Tennant, C., & Blignault, I. (1985). Predicting improvement in patients with non-endogenous depression. *British Journal of Psychiatry, 146*, 132–139.

14. Tang, T. Z., Luborsky, L., & Andrusyna, T. (2002). Sudden gains in recovering from depression: Are they also found in psychotherapies other than cognitive-behavioral therapy? *Journal of Consulting and Clinical Psychology, 70*, 444–447; Tang, T. Z., DeRubeis, R. J., Hollon, S. D., Amsterdam, J., & Shelton, R. (2007). Sudden gains in cognitive therapy of depression and depression relapse/recurrence. *Journal of Consulting and Clinical Psychology, 75*, 404–408.

15. Clearly some people err when they attribute rapid improvement to treatment.

16. Most studies of early improvement rely on treated samples. Many people who improve early and stay well never present for treatment and thus leave little trace in the research literature; we know very little about rapid improvement in people who are never treated.

17. These problems are maintaining factors for depression, and the obverse is true: depression can be a maintaining factor for these other problems.

18. Viinamäki, H., Tanskanen, A., Honkalampi, K., Koivumaa-Honkanen, H., Antikainen, R., Haatainen, K., & Hintikka, J. (2006). Recovery from depression: A two-year follow-up study of general population subjects. *International Journal of Social Psychiatry, 52*, 19–28.

19. Related research has shown that a number of stressful life events have nonlinear effects on depression incidence. Kendler, K. S., Karkowski, L. M., & Prescott, C. A. (1998). Stressful life events and major depression: Risk period, long-term contextual threat, and diagnostic specificity. *Journal of Nervous and Mental Disease, 186*, 661–669.

20. Geschwind, N., Nicolson, N. A., Peeters, F., van Os, J., Barge-Schaapveld, D., & Wichers, M. (2011). Early improvement in positive rather than negative emotion predicts remission from depression after pharmacotherapy. *European Neuropsychopharmacology, 21,* 241–247.

21. Leenstra, A. S., Ormel, J., & Giel, R. (1995). Positive life change and recovery from depression and anxiety. A three-stage longitudinal study of primary care attenders. *British Journal of Psychiatry, 166,* 333–343.

22. Brown, G. W., Adler, Z., & Bifulco, A. (1988). Life events, difficulties and recovery from chronic depression. *British Journal of Psychiatry, 152,* 487–498; Harris, T., Brown, G. W., & Robinson, R. (1999). Befriending as an intervention for chronic depression among women in an inner city. 2: Role of fresh-start experiences and baseline psychosocial factors in remission from depression. *British Journal of Psychiatry, 174,* 225–232. Interestingly, in some samples of severely depressed persons, negative life events do not appear to adversely affect course. Paykel, E. S., Cooper, Z., Ramana, R., & Hayhurst, H. (1996). Life events, social support and marital relationships in the outcome of severe depression. *Psychological Medicine, 26,* 121–133.

23. Paykel, E. S. (2003). Life events and affective disorders. *Acta Psychiatrica Scandinavica, 108,* 61–66. See also Oldehinkel, A. J., Ormel, J., & Neeleman, J. (2000). Predictors of time to remission from depression in primary care patients: Do some people benefit more from positive life change than others? *Journal of Abnormal Psychology, 109,* 299–307. See also Needles, D. J., & Abramson, L. Y. (1990). Positive life events, attributional style, and hopefulness: Testing a model of recovery from depression. *Journal of Abnormal Psychology, 99,* 156–165; Overbeek, G., Vermulst, A., de Graaf, R., ten Have, M., Engels, R., & Scholte, R. (2010). Positive life events and mood disorders: Longitudinal evidence for an erratic lifecourse hypothesis. *Journal of Psychiatric Research, 44,* 1095–1100.

24. I doubt I will ever have a complete fix on why I improved. Having depression and studying mood can be equally humbling!

25. Judd, L. L., Akiskal, H. S., Zeller, P. J., Paulus, M., Leon, A. C., Maser, J. D., & Keller M. B. (2000). Psychosocial disability during the long-term course of unipolar major depressive disorder. *Archives of General Psychiatry, 57,* 375–380. See also McKnight, P. E., & Kashdan, T. B. (2009). The importance of functional impairment to mental health outcomes: A case for reassessing our goals in depression treatment research. *Clinical Psychology Review, 29,* 243–259.

26. Iacoviello, B. M., Alloy, L. B., Abramson, L. Y., & Choi, J. Y. (2010). The early course of depression: A longitudinal investigation of prodromal symptoms and their relation to the symptomatic course of depressive episodes. *Journal of Abnormal Psychology, 119,* 459–467.

27. Fava, G. A., Grandi, S., Zielenzy, M., Canestrari, R., & Morphy, M. A. (1994). Cognitive behavioral treatment of residual symptoms in primary major depressive disorder. *American Journal of Psychiatry, 151,* 1295–1299;

Mahnert, F. A., Reicher, H., Zalaudek, K., & Zapotoczky, H. G. (1997). Prodromal and residual symptoms in recurrent depression: Preliminary data of a long-term study under prophylactic treatment condition. *European Neuropsychopharmacology*, *7*, s159–s160.

28. Keller, M. B., Lavori, P. W., Mueller, T. I., Endicott, J., Coryell, W., Hirschfeld, R. M. A., & Shea, T. (1992). Time to recovery, chronicity, and levels of psychopathology in major depression: A 5-year prospective follow-up of 431 subjects. *Archives of General Psychiatry*, *49*, 809–816.

29. Episodes longer than two years are often considered the boundary for chronic depression. Chronic depression may be found in 20 percent or more of participants in clinical samples and is considerably lower in community samples.

30. Quoted in Riso, L. P., Miyatake, R. K., & Thase, M. E. (2002). The search for determinants of chronic depression: a review of six factors, *Journal of Affective Disorders*, *70*, 103–115. People who have shown early signs of chronicity are more likely to have chronic courses of depression. Klein, D. N., Norden, K. A., Ferro, T., Leader, J. B., Kasch, K. L., Klein, L. M., & Aronson, T. A. (1998). Thirty-month naturalistic follow-up study of early-onset dysthymic disorder: Course, diagnostic stability, and prediction of outcome. *Journal of Abnormal Psychology*, *107*, 338–348.

31. Denton, W. H., Carmody, T. J., Rush, A. J., Thase, M. E., Trivedi, M. H., Arnow, B. A., & Keller, M. B. (2010). Dyadic discord at baseline is associated with lack of remission in the acute treatment of chronic depression. *Psychological Medicine: A Journal of Research in Psychiatry and the Allied Sciences*, *40*, 415–424; Brown, G. W., Harris, T. O., Kendrick, T., Chatwin, J., Craig, T. K. J., Kelly, V., & the Thread Study Group. (2010). Antidepressants, social adversity and outcome of depression in general practice. *Journal of Affective Disorders*, *121*, 239–246.

32. Monroe, S. M., Kupfer, D. J., & Frank, E. (1992). Life stress and treatment course of recurrent depression: I. Response during index episode. *Journal of Consulting and Clinical Psychology*, *60*, 718–724.

33. Heffernan, V. (2001). A delicious placebo. In N. Casey (Ed.), *Unholy ghost: Writers on depression* (pp. 8–20). New York: William Morrow.

34. Again, our advanced language capability makes it easy for the experience of depression to be fed forward as a further input into the mood system. Over time, the person comes to represent himself or herself as a "depressive" who "expects" to be depressed.

35. Styron, W. (2010). *Darkness visible: A memoir of madness*. New York: Open Road Media.

Chapter Nine: In Limbo

1. Richey, C., & Kallendorf, H. R. (2010). *Acing depression: A tennis champion's toughest match*. Washington, DC: New Chapter Press, p. 215.

2. Thompson, T. (1996). *The beast: A journey through depression*. New York: Plume.

3. Most algorithms and treatment protocols are designed for acute depression.

4. Low-grade symptoms are such a common outcome that some have even asked if complete and total remission is a realistic goal. Keitner, G. I., Solomon, D. A., & Ryan, C. E. (2008). STAR*D: Have we learned the right lessons? *American Journal of Psychiatry, 165,* 133.

5. It's telling that research studies tend not to break out the residual symptoms of depression individually. For interesting exceptions, see Minor, K. L., Champion, J. E., & Gotlib, I. H. (2005). Stability of DSM-IV criterion symptoms for major depressive disorder. *Journal of Psychiatric Research, 39,* 415–420; and Conradi, J., Ormel, J., & de Jonge, P. (2011). Presence of individual (residual) symptoms during depressive episodes and periods of remission: A 3-year prospective study. *Psychological Medicine, 41,* 1165–1174.

6. Nierenberg, A. A., Keefe, B. R., Leslie, V. C., Alpert, J. E., Pava, J. A., Worthington, J. J. III, . . . Fava M. (1999). Residual symptoms in depressed patients who respond acutely to fluoxetine. *Journal of Clinical Psychiatry, 60,* 221–225.

7. Paykel, E. S., Ramana, R., Cooper, Z., Hayhurst, H., Kerr, J., & Barocka, A. (1995). Residual symptoms after partial remission: An important outcome in depression. *Psychological Medicine, 25,* 1171–1180.

8. Judd, L. L., Akiskal, H. S., Maser, J. D., Zeller, P. J., Endicott, J., Coryell, W., . . . Keller, M. B. (1998). Major depressive disorder: A prospective study of residual subthreshold depressive symptoms as predictor of rapid relapse. *Journal of Affective Disorders, 50,* 97–108. The difference was twenty-three weeks versus sixty-eight weeks.

9. Judd, L. L., Paulus, M. J., Schettler, P. J., Akiskal, H. S., Endicott, J., Leon, A. C., . . . Keller, M. B. (2000). Does incomplete recovery from first lifetime major depressive episode herald a chronic course of illness? *American Journal of Psychiatry, 157,* 1501–1504.

10. Because of the general lack of focus on residual depression, less thought has gone into how low-grade depression should be assessed.

11. The life stress literature has shown that major life stress is more closely associated with the first episode than with later episodes of depression. Monroe, S. M., & Harkness, K. L. (2005). Life stress, the "kindling" hypothesis, and the recurrence of depression: Considerations from a life stress perspective. *Psychological Review, 112,* 417–445.

12. Post, R. M. (1992). Transduction of psychosocial stress into the neurobiology of recurrent affective disorder. *American Journal of Psychiatry, 149,* 999–1010.

13. This analogy is somewhat ironic, because inducing seizures via electroconvulsive therapy is a recognized means of treating an episode of severe depression.

14. See Pinker, S. (1997). *How the mind works*. New York: W. W. Norton for a great discussion of how affective memories are critical for guiding future behavior. Despite compelling converging logic that mood congruency reflects evolved design, it is difficult to prove.

15. Teasdale, J. D., & Fogarty. S. J. (1979). Differential effects of induced mood on retrieval of pleasant and unpleasant events from episodic memory. *Journal of Abnormal Psychology*, *88*, 248–257.

16. Rottenberg, J., Hildner, J. C., & Gotlib, I. H. (2006). Idiographic autobiographical memories in major depressive disorder. *Cognition & Emotion*, *20*, 114–128.

17. Joormann, J., Siemer, M., & Gotlib, I. H. (2007). Mood regulation in depression: Differential effects of distraction and recall of happy memories on sad mood. *Journal of Abnormal Psychology*, *116*, 484–490.

18. Kenneth Malmberg, personal communication with the author, January 31, 2012.

19. Coyne, J. C., & Gotlib, I. H. (1986). Studying the role of cognition in depression: Well-trodden paths and cul-de-sacs. *Cognitive Therapy and Research*, *10*, 695–705.

20. Segal, Z. V., & Ingram, R. E. (1994). Mood priming and construct activation in tests of cognitive vulnerability to unipolar depression. *Clinical Psychology Review*, *14*, 663–695.

21. Gary Marcus writes, "In many circumstances, especially those requiring snap decisions, recency, frequency, and context are powerful tools for mediating memory. For our ancestors, who lived almost entirely in the here and now (as virtually all nonhuman life forms still do), quick access to contextually relevant memories of recent events or frequently occurring ones helped navigate the challenges of seeking food or avoiding danger." Marcus, G. (2008). *Kluge: The haphazard construction of the human mind*. New York: Houghton Mifflin, pp. 36–37.

22. Miranda, J., & Persons, J. B. (1988). Dysfunctional attitudes are mood-state dependent. *Journal of Abnormal Psychology*, *97*, 76–79; Miranda, J., Gross, J. J., Persons, J. B., & Hahn, J. (1998). Mood matters: Negative mood induction activates dysfunctional attitudes in women vulnerable to depression. *Cognitive Therapy and Research*, *22*, 363–376. See also Teasdale, J. D. (1988). Cognitive vulnerability to persistent depression. *Cognition and Emotion*, *2*, 247–274.

23. Segal, Z. V., Gemar, M., & Williams, S. (1999). Differential cognitive response to a mood challenge following successful cognitive therapy or pharmacotherapy for unipolar depression. *Journal of Abnormal Psychology*, *108*, 3–10. The patients had to have been in remission for at least ten weeks.

24. Greden, J. F. (Ed.). (2001). *Treatment of recurrent depression*. Review of psychiatry series, Vol. 20. Washington, DC: American Psychiatric Publishing, p. 5.

25. Segal, Z. V., Bieling, P., Young, T., MacQueen, G., Cooke, R., Martin, L., Bloch, R., & Levitan, R. D. (2010). Antidepressant monotherapy vs. sequential pharmacotherapy and mindfulness-based cognitive therapy, or placebo, for relapse prophylaxis in recurrent depression. *Archives of General Psychiatry, 67,* 1256–1264; Bockting, C. L. H., Schene, A. H., Spinhoven, P., Koeter, M. W. J., Wouters, L. F., Huyser, J. & Kamphuis, J. H. (2005). Preventing relapse/recurrence in recurrent depression with cognitive therapy: A randomized controlled trial. *Journal of Consulting and Clinical Psychology, 73,* 647–657. There is also evidence for the potentially exciting claim that a course of cognitive therapy with only minimal follow-up may have an enduring effect and prevent relapse. Hollon, S. D., DeRubeis, R. J., Shelton, R. C., Amsterdam, J. D., Salomon, R. M., O'Reardon, J. P., & Gallop, R. (2005). Prevention of relapse following cognitive therapy vs. medications in moderate to severe depression. *Archives of General Psychiatry, 62,* 417–422.

26. Monroe S. M., & Harkness, K. L. (2011). Recurrence in major depression: A conceptual analysis. *Psychological Review, 118,* 655–674.

27. I saw some top psychiatrists in affective disorders. They did their level best to make me better.

28. A positive family history of depression renders a person more vulnerable to depression. But that's also true of a positive family history of anxiety, substance problems, and schizophrenia.

29. Monroe S. M., & Harkness, K. L. (2011). Recurrence in major depression: A conceptual analysis. *Psychological Review, 118,* 655–674.

Chapter Ten: The Glory of Recovery

1. Monroe, S. M., & Harkness, K. L. (2011). Recurrence in major depression: A conceptual analysis. *Psychological Review, 118,* 655–674. They estimate that 40 to 50 percent of people with a first episode of depression will never have another.

2. The genre of self-help for depression can be guilty of being overly glib. I remember well my own wanderings in the bookstore, looking at shelves full of bright titles about how one could vanquish one's depression in ten easy steps.

3. See Carver, C. S. (1998). Resilience and thriving: Issues, models, and linkages. *Journal of Social Issues, 54,* 245–266.

4. These three areas are discussed separately, but they are interconnected. For example, exercises that might increase personal growth also increase well-being. Seligman and colleagues did a randomized controlled trial on the Internet to demonstrate the possibility of improving well-being in ordinary people with short psychologically-based exercises. See Seligman, M. E., Steen, T. A., Park, N., & Peterson, C. (2005). Positive psychology progress: Empirical validation of interventions. *American Psychologist, 60,* 410–421.

5. On March 7, 2012, I sent out a challenge on my *Psychology Today* blog to identify and point out even a single research study on the topic of flourishing after depression. This post has been viewed more than two thousand times, and thus far no one has nominated a research study. See https://my .psychologytoday.com/blog/charting-the-depths/201203/flourishing -after-depression-how-what-we-dont-know-hurts-us. Similarly, Monroe and Harkness (see note 1) write about the lack of research on people who have a single lifetime episode of depression (the SLEDs): "But because these single lifetime cases have fallen between the cracks of theory, research, and practice, it is not surprising that almost nothing is known about them. This class of formerly depressed persons requires detailed attention and analysis."

6. Wood, A. M., & Joseph, S. (2010). The absence of positive psychological (eudemonic) well-being as a risk factor for depression: A ten-year cohort study. *Journal of Affective Disorders, 122,* 213–217.

7. Outside of depression, a rich vein of research considers the ingredients of human flourishing, sometimes referred to as "positive psychology." Building on pioneering work by people like Rogers and Maslow, contemporary positive psychologists have enhanced our understanding of how, why, and under what conditions people are able to flourish. Sheldon, K., Kashdan, T. B., & Steger, M. F. (2011). (Eds.). *Designing positive psychology: Taking stock and moving forward.* New York: Oxford University Press.

8. A good popular summary is Joseph, S. (2011*). What doesn't kill us: The new psychology of posttraumatic growth.* New York: Basic Books.

9. Rendon, J. (2012, March 22). Post-traumatic stress's surprisingly positive flip side. *New York Times.* Retrieved from http://www.nytimes .com/2012/03/25/magazine/post-traumatic-stresss-surprisingly-positive -flip-side.html?pagewanted=all&_r=0; Tedeschi, R. G., Park, C. L., & Calhoun, L. G. (Eds.). (1998). *Posttraumatic growth: Positive changes in the aftermath of crisis.* Mahwah, NJ: Lawrence Erlbaum Associates.

10. Flach's *The Secret Strength of Depression* is a fine discussion of the ways that depressions present opportunities for personal growth. Flach, F. (2002). *The secret strength of depression* (3rd ed.). Hobart, NY: Hatherleigh Press. Unfortunately few empirical researchers have picked up on his insights.

11. See Lyubomirsky, S., King, L., & Diener, E. (2005). The benefits of frequent positive affect: Does happiness lead to success? *Psychological Bulletin, 131,* 803–855; Lyubomirsky, S., Sheldon, K. M., & Schkade, D. (2005). Pursuing happiness: The architecture of sustainable change. *Review of General Psychology, 9,* 111–131.

12. Harker, L., & Keltner, D. (2001). Expressions of positive emotion in women's college yearbook pictures and their relationship to personality and life outcomes across adulthood. *Journal of Personality and Social Psychology, 80,* 112–124; Okun, M. A., Stock, W. A., Haring, M. J., & Witter,

R. A. (1984). The social activity/subjective well-being relation: A quantitative synthesis. *Research on Aging, 6*, 45–65.

13. Estrada, C. A., Isen, A. M., & Young, M. J. (1994). Positive affect improves creative problem solving and influences reported source of practice satisfaction in physicians. *Motivation and Emotion, 18*, 285–299; Staw, B. M., Sutton, R. I., & Pelled, L. H. (1994). Employee positive emotion and favorable outcomes at the workplace. *Organization Science, 5*, 51–71.

14. Fredrickson, B. L. (1998). What good are positive emotions? *Review of General Psychology, 2*, 300–319. For application of this perspective to psychological problems, see Garland, E. L., Fredrickson, B., Kring, A. M., Johnson, D. P., Meyer, P. S., & Penn, D. L. (2010). Upward spirals of positive emotions counter downward spirals of negativity: Insights from the broaden-and-build theory and affective neuroscience on the treatment of emotion dysfunctions and deficits in psychopathology. *Clinical Psychology Review, 30*, 849–864.

15. Cohn, M. A., Fredrickson, B. L., Brown, S. L., Mikels, J. A., & Conway, A. M. (2009). Happiness unpacked: Positive emotions increase life satisfaction by building resilience. *Emotion, 9*, 361–368.

16. John Lennon, "Beautiful Boy" (from the last album of this British singer/songwriter, 1940–1980).

17. See review article by McKnight, P. E., & Kashdan, T. B. (2009). Purpose in life as a system that creates and sustains health and well-being: An integrative, testable theory. *Review of General Psychology, 13*, 242–251.

18. Existentialist perspectives on depression merit serious consideration. Maisel, E. (2012). *Rethinking depression: How to shed mental health labels and create personal meaning*. Novato, CA: New World Library. A great effort to synthesize contemporary psychiatry with an existentialist perspective is Ghaemi, N. (2013). *On depression: Drugs, diagnosis, and despair in the modern world*. Baltimore: Johns Hopkins University Press.

19. For an exception, see Gary Greenberg's wonderfully provocative book. Greenberg, G. (2010). *Manufacturing depression: The secret history of a modern disease*. New York: Simon & Schuster.

20. Kirsch, I. (2010). *The emperor's new drugs: Exploding the antidepressant myth*. New York: Basic Books.

Recommended Readings

Evolution and Evolutionary Approaches to Mental Disorders

Nesse, R. M. (2000). Is depression an adaptation? *Archives of General Psychiatry, 57,* 14–20.

Nesse, R. M., & Williams, G. C. (1996). *Why we get sick: The new science of Darwinian medicine.* New York: Vintage.

Pinker, S. (1997). *How the mind works.* New York: W. W. Norton & Company.

Wilson, D. S. (2007). *Evolution for everyone: How Darwin's theory can change the way we think about our lives.* New York: Bantam Dell.

Science of Mood and Emotion

Gross, J. J. (Ed.). (2009). *Handbook of emotion regulation.* New York: Guilford Press.

Gruber, J., Mauss I. B., & Tamir M. (2011). A dark side of happiness? How, when, and why happiness is not always good. *Perspectives on Psychological Science, 6,* 222–233.

Kring, A. M., & Sloan, D. M. (Eds.). (2010). *Emotion regulation and psychopathology: A transdiagnostic approach to etiology and treatment.* New York: Guilford Press.

Rottenberg, J. E., & Johnson, S. L. (Eds.). (2007). *Emotion and psychopathology: Bridging affective and clinical science.* Washington, DC: American Psychological Association.

Animal Emotions

Balcombe, J. (2006). *Pleasurable kingdom: Animals and the nature of feeling good.* New York: Macmillan.

King, B. J. (2013). *How animals grieve*. Chicago: University of Chicago Press.

Panksepp, J., & Biven, L. (2012). *The archaeology of mind: Neuroevolutionary origins of human emotions*. New York: W. W. Norton & Company.

Grief

Archer, J. (2004). *The nature of grief: The evolution and psychology of reactions to loss*. London: Routledge.

Bonanno, G. A. (2009). *The other side of sadness: What the new science of bereavement tells us about life after loss*. New York: Basic Books.

Mental Illness and Psychiatric Diagnosis

McNally, R. J. (2011). *What is mental illness?* Cambridge, MA: Harvard University Press.

Watters, E. (2010). *Crazy like us: The globalization of the American psyche*. New York: Free Press.

Depression and Depression Treatment

Andrews, P. W., & Thomson, J. A., Jr. (2009). The bright side of being blue: Depression as an adaptation for analyzing complex problems. *Psychological Review, 116*, 620–654.

Ghaemi, N. (2013). *On depression: Drugs, diagnosis, and despair in the modern world*. Baltimore, MD: Johns Hopkins University Press.

Greenberg, G. (2010). *Manufacturing depression: The secret history of a modern disease*. New York: Simon & Schuster.

Healy, D. (1999). *The antidepressant era*. Cambridge, MA: Harvard University Press.

Horwitz, A. V., & Wakefield, J. C. (2007). *The loss of sadness: How psychiatry transformed normal sorrow into depressive disorder*. New York: Oxford University Press.

Jamison, K. R. (1995). *An unquiet mind: A memoir of moods and madness*. New York: Alfred A. Knopf.

Karp, D. A. (1997). *Speaking of sadness: Depression, disconnection, and the meanings of illness*. New York: Oxford University Press.

Kirsch, I. (2011). *The emperor's new drugs: Exploding the antidepressant myth*. New York: Basic Books.

Solomon, A. (2002). *The noonday demon: An atlas of depression*. New York: Scribner.

Self-Help and Happiness

Gilbert, D. (2006). *Stumbling on happiness.* New York: Vintage.

Harris, R. (2008). *The happiness trap: How to stop struggling and start living.* Boston: Shambhala Publications.

Lyubomirsky, S. (2013). *The myths of happiness: What should make you happy, but doesn't, what shouldn't make you happy, but does.* New York: Penguin.

Maisel, E. (2012). *Rethinking depression: How to shed mental health labels and create personal meaning.* Novato, CA: New World Library.

Index